NANOELECTRONICS AND MATERIALS DEVELOPMENT

Edited by **Abhijit Kar**

Nanoelectronics and Materials Development

http://dx.doi.org/10.5772/61560
Edited by Abhijit Kar

Contributors

Yoshio Ichida, Fei Song, Kongchao Shen, Yingguo Yang, Fan Wu, Nan Yao, Hui Li, Yi Zhou, Jichen Dong, Nurul Amziah Md Yunus, Nima Khoshsirat, Rafael Vargas-Bernal, Deepak Rawtani

CBS, Edition **2017**

Published by InTech

Janeza Trdine 9, 51000 Rijeka, Croatia

Publishing Process Manager Dajana Pemac

Technical Editor SPi Global

Cover Designer InTech Design Team

Additional hard copies can be obtained from orders@intechopen.com

Nanoelectronics and Materials Development, Edited by Abhijit Kar
p. cm.
Print ISBN 978-953-51-2525-9
Online ISBN 978-953-51-2526-6

INTECH

open science | open minds

Contents

Preface

Nanoelectronics mainly targets to discourse the direct crisis of Moore's law by developing new materials in nanoscales. Since Moor's observation, during the last few decades, the sizes of transistor have decreased from 10 μm to 22 nm range. In the twenty-first century, the expectation is to analyze the actions of atoms using a computer program. It is based on the original Planck electronic model of the atom which has been neglected and forgotten by most, for quite a long time. Application of electronic materials is highly observed in almost all industries and all areas of the society. In recent past, serious environmental hazards caused out of the enormous use of electronic gadgets containing lead have been observed. As a result, the electronics industry is migrating to lead-free electronics driven by impending government legislation and market forces. Recently it has become the most challenging task to develop the miniaturized reliable lead-free electronic devices.

Now it is realized that the issues of sustainable development and transformational research practice should be given the highest priority in order to combat the futuristic scientific challenges and demand. This book has discussed some of the most advanced research observations in very concise and structured way which is indispensable toward nanomaterials development and device-making.

Seven different chapters in this book, namely, (i) *State-of-the-Art of Electronic Devices Based on Graphene;* (ii) *Aspects of Nanoelectronics in Materials Development;* (iii) *Copper Indium Gallium diSelenide (CIGS) Nanocrystalline Bulk Semiconductor as the Absorber Layer, Its Current Technological Trends, and Optimization;* (iv) *Epitaxial Cu3Ge Thin Film: Fabrication, Structure, and Property;* (v) *Nanomachining of Fused Quartz Using Atomic Force Microscopy;* (vi) *Fabrication and Characterization of Organic-Inorganic Hybrid Perovskite Devices with External Doping;* (vii) *First-Principle Study of the Electron Transport Properties of Graphene-Like 2D Materials,* are written by very well-acclaimed scientists mostly from the United States, China, Hong Kong, Australia, Mexico, Japan, and India rendering necessary facts on all aspects of nanotechnology, its application, and immense possibilities of advanced materials development.

The main focus of the book was dragged toward those materials where electronic properties are manipulated for development of advanced materials. We have discussed about the extensive usage of nanotechnology and its impact on various facets of the chip-making practice from materials to devices like basic memory, quantum dots, nanotubes, nanowires, graphene-like 2D materials, and CIGS thin-film solar cells as energy-harvesting devices.

Finally, I must mention that it was a wonderful experience of working throughout the publication process with InTechOpen publishing and my publishing team members. Also I must thank SERB-DST and Government of India for their support. My sincere thanks go to the director and my colleagues at Jagadis Bose National Science Talent Search, Kolkata, India, for their valuable feedback.

And last but not the least, the intention of writing this book was to present in a concise form various exciting and interesting studies on nanotechnology and its applications for futuristic materials development, carried out by scientists from different parts of the globe. Researchers and students can gain valuable insights into the different processing of nanomaterials, characterization procedures of the materials in nanoscale, and their different functional properties and applications. I hope that the book will open out a new vista in the field of transformational research on nanotechnology.

Dr. Abhijit Kar
Jagadis Bose National Science Talent Search
Kolkata, India

Chapter 1

State-of-the-Art Electronic Devices Based on Graphene

Rafael Vargas-Bernal

Additional information is available at the end of the chapter

http://dx.doi.org/10.5772/64320

Abstract

Graphene can be considered as the material used for electronic devices of this century, due to its excellent physical and chemical properties, which have been studied and implemented from a theoretical basis and have allowed the development of unique and innovative applications. The need for an ongoing study of the state-of-the-art electronic devices is ultimately useful for the progress achieved so far and future project applications. To date, graphene has been used individually in composite, hybrid materials or functional materials. In this chapter, an overview of their applications in nanoelectronics, particularly with an emphasis directed to flexible electronics, is presented. The description of the advantages and properties of graphene at a level of materials science and engineering is presented, in order to spread its enormous potential. In addition, the future prospects of these applications arising from the developments made currently in the laboratory phase are examined.

Keywords: graphene, nanoelectronics, flexible electronics, electronic devices

1. Introduction

The main driving force of the electronics industry is the search of new materials, capable of fulfilling the compelling demand for a higher performance and lower power consumption in the electronic systems. Novel electronic devices based on two-dimensional materials are being designed as innovations for flexible electronics within new perspectives of the future technological developments [1, 2]. Numerous research groups around the world are introducing nanomaterials which can work individually, or used in combination with other materials to exploit the physicochemical properties of these materials either as composite materials, hybrid materials, or functional materials. In particular, carbon nanomaterials such as carbon nanotubes and graphene are impelling the innovation in the area of electronics through diverse devices making use of different technological strategies by exploiting the materials science and engineering.

Among the allotropes of carbon, graphene offers one of the best materials to develop applications in areas such as electronics, biological engineering, filtration, lightweight and strong composites and photovoltaic and energy storage applications [3, 4]. Since the isolation of graphene from graphite in 2004 by Andre Geim and Konstantin Novoselov at the University of Manchester, this electronic material has gained considerable interest in different fields of application in the last decade [2, 5]. Its strategic advantages are derived from the mechanical, chemical, electronic, optical, thermal, magnetic and biological properties. This material is 207 times stronger than steel by weight, conducts heat and electricity efficiently and is almost transparent. Graphene is an emerging material for future electronics directed into flexible electronics, photonics and electrochemical energy storage [6], as shown in **Figure 1**.

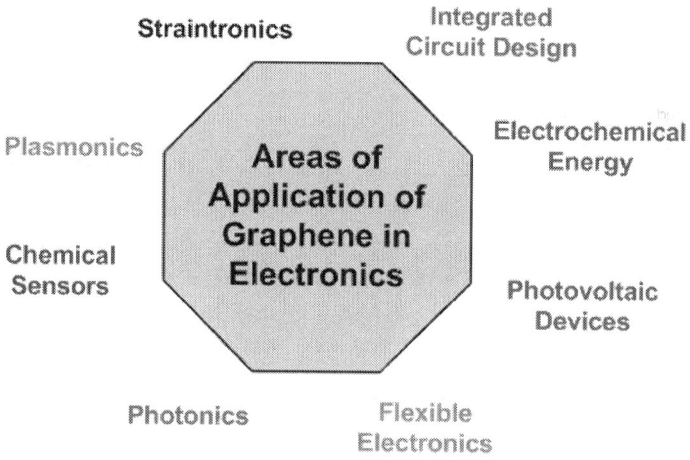

Figure 1. Technical areas of application of the graphene in electronics industry.

Different authors have published studies about the state-of-the-art graphene and its applications [4, 7]; however, it is impossible that all varieties of applications and innovations achieved to date can be covered in a unique work. In flexible nanoelectronics, graphene is primarily used in RF FETs, transparent conductive films, heat spreaders, acoustic speakers and mechanical actuators [7]. Commercial products bearing graphene are touch panels of smartphones by companies such as Samsung, Nokia and Sony. For example, hybrid materials have extended functionalities of graphene in different applications such as resonant tunnelling devices, light emission devices, photovoltaic devices, plasmonics, chemical sensors including gas sensors and flexible electronics [6], as shown in **Figure 2**. In this chapter, the main advantages of graphene in the electronics industry are analysed through their various technological applications. A brief description collecting relevant information about graphene and its applications is presented to summarize its extraordinary potential. A comprehensive review of the progress made and reported in the literature in the last decade is performed in order to predict its future applications.

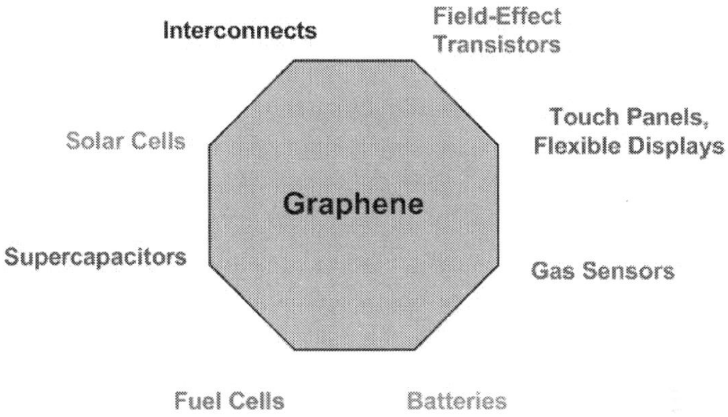

Figure 2. Main electronic devices fabricated based on graphene of the electronics industry.

2. Electrical properties of the graphene and basic devices

Graphene can be defined as a two-dimensional crystalline material composed of a graphite monolayer with a thickness of 0.34 nm, where carbon atoms present a sp^2 hybridization state since each atom is covalently bonded to three others and these form a honeycomb lattice composed of two intertwined triangular sublattices, as shown in **Figure 3**. Its mechanical properties such as extraordinary strength and flexibility are derived from the strong and rigid σ bonds formed by its sp^2 hybrid orbitals, while electrical properties are obtained due to the localization of the σ bonds and formation of π and π^* bonds by hybridization of the remaining p_z atomic orbitals of the nearest C atoms, thereby making that the electrons behave as a 2D electron gas [8–10]. Graphene has the ability to accept electrons from and donate to the strong electron donors or acceptors, respectively. These charge-transfer processes will lead to *n*- or *p*-doping of the graphene conducting to partially charged species at facilitating electrostatic interactions between them. Graphene can also establish strong van der Waals and π–π interactions with other moieties due to its insolubility and extreme aspect ratio [11]. Therefore, graphene has an enormous capability of adsorbing small molecules and therefore, it is extremely sensible to be used as a sensing material, although it also presents a very poor selectivity [10]. Graphene and its derivatives can react with a wide variety of chemical substances. These reactions, for example, chemical functionalization, are used to modulate the structures and properties of the graphene with the aim of extending their functionalities and practical applications [12]. Graphene functionalization is carried out either in a noncovalent or covalent manner. Weak interactions of the type π-π, van der Waals or electrostatic are observed in noncovalent functionalization, while an oxygen-containing functional group (carboxylic, epoxic and/or hydroxyl) is produced in covalent functionalization [13]. Chemical doping of the graphene facilitates the tuning of the electronic structure and properties by changing their electrical properties from metallic to semiconducting behaviour [14]. Dopants present at the interstitial site in graphene can be removed by a suitable heat treatment. When graphene is doped with heteroatoms (for example, N, B, P or S), more active sites are produced and its electronic properties are tuned, thus improving the interactions between graphene and oxygen molecules [15].

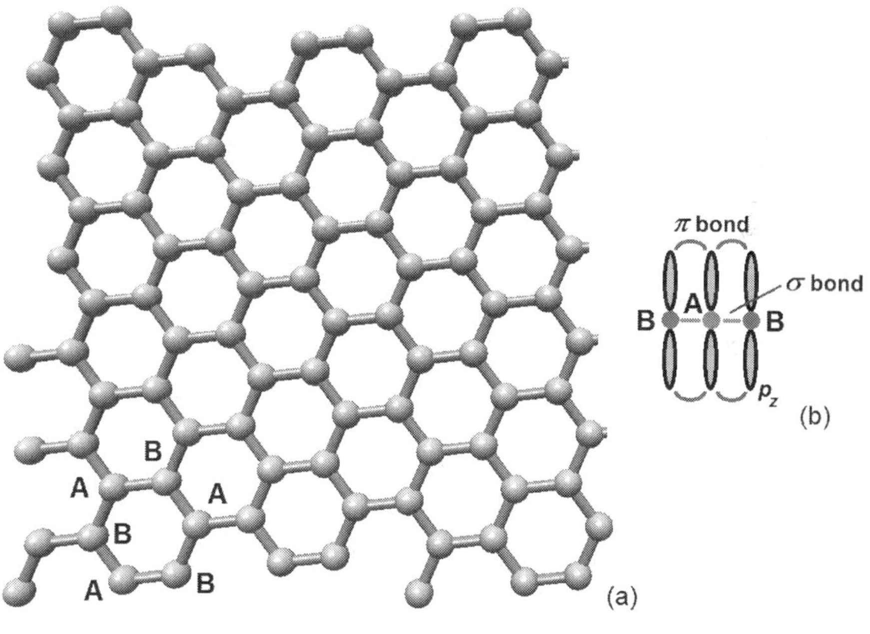

Figure 3. Basic aspects of graphene: (a) a sheet of graphene, and (b) types of chemical bonds presented in graphene.

Like carbon nanotubes, graphene has impressive electrical transport properties. Each intertwined triangular sub-lattice of the honeycomb lattice contributes to the wavefunction of charge carriers. Its unique conduction properties can be described by an energy dispersion equation, Eq. (1), which leads to the vanishing of the energy bandgap in the so-called Dirac points illustrated in **Figure 4**. The energy dispersion can be expressed as follows [9, 16]:

$$E(\boldsymbol{k}) = \pm t\sqrt{3 + 2cos\left(\sqrt{3}k_y a\right) + 4cos\left(\frac{\sqrt{3}k_y a}{2}\right)cos\left(\frac{\sqrt{3}k_x a}{2}\right)} \tag{1}$$

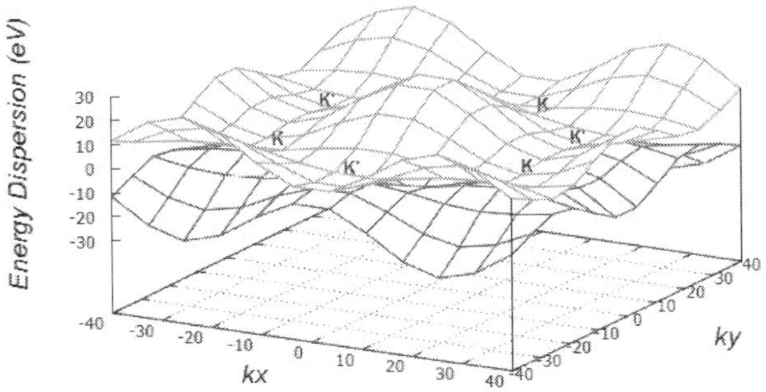

Figure 4. Energy dispersion in graphene.

where $\boldsymbol{k} = (k_x, k_y)$ is the wavevector of charge carriers relative to Dirac points, $a = 0.142$ nm is the distance between two C atoms, $t = 2.75$ eV is known as the hopping energy, plus and minus signs are associated with the upper (π^*) and lower (π) bands, which are referred as the electron and hole states, respectively [17]. The honeycomb crystal structure of single-layer graphene consists of two non-equivalent sublattices and results in a unique band structure for the itinerant π-electrons near the Fermi energy which behave as massless Dirac fermion. The valence and conduction bands touch conically at two non-equivalent Dirac points, called K and K' point, which form a time-reserved pair, i.e. opposite chirality [18, 19]. Moreover, electronic properties of graphene are invariant by interchanging the K and K' states, which means that the two valleys are related by time-reversal symmetry [8]. It can be observed in **Figure 2** that the electrical transport properties are equal for electron states or hole states due to the symmetry around the Dirac points.

The electrical transport in graphene is ambipolar, that is, it can be developed by electrons or holes, depending on the electrical voltage applied to the material either positive or negative, respectively. Both ambipolar transport and the lack of a bandgap at Dirac points have conducted to the so-called Klein paradox (Klein tunnelling), which implies that charge carrier transport leads to the complete transformation of electron states into hole states (or vice versa) [9, 17, 20]. The Klein paradox implies that the reflected electrical current is larger than the incident one and the reflection probability is larger than unity [11, 21].

Intrinsic graphene is a semimetal or zero-gap semiconductor. Monolayer graphene has a cone-like valence and conduction bands intersect at the Fermi level with no band gap, even a bilayer graphene without electrical field applied has the behaviour of the gapless semiconductor. Bilayer graphene shows a band gap when an electrical field is applied in a direction perpendicular to the σ-bond [22]. Graphene contains 100% sp^2 orbitals; however, if some of these orbitals are converted to sp^3 orbitals, then it presents a band gap and a semiconducting behaviour can be exploited. When graphene is subjected to twisting, it strongly affects the band structures of graphene, and electron localization is modified, and it changes the nature and magnitude of the electrical current passing through graphene.

Graphene conducts either electrons or holes with concentration as high as 10^{13} cm^{-3}. It has an extraordinary carrier mobility of ~500,000 cm^2/Vs and its electronic properties are strongly related to its thickness [22]. Due to its high electrical and thermal conductivity (5000 Wm/K) and low electrical noise, graphene is considered as an interesting alternative to copper for electrical interconnects in integrated circuits to connect electronic devices [23, 24]. Vertical and horizontal interconnections can be implemented using zigzag graphene nanoribbons, where horizontal connections are more feasible. The graphene presents a higher conductance with respect to Cu for interconnects in the range of nanometers. The following properties of graphene have been exploited in interconnects: high carrier mobility at room temperature, thermal conductivity, higher mechanical strength, reduced capacitance coupling between adjacent wires, width-dependent transport gap, temperature coefficient and ballistic transport. Graphene shows its work function dependence on the type of substrate used for its synthesis. It has a very large surface area 2630 m^2/g. Graphene has to be chemically modified according to the application as well as the design of different electronic devices. The electrical mobility in graphene depends completely on the physical properties of the substrate on which this material is deposited to be used in electronic applications [9, 11]. Particularly, this parameter establishes the performance that, for example, graphene-based field-effect transistors (GFETs) will achieve [25]. In addition, configurations such as those based on top gate (TG) and where materials for oxide with high dielectric constants (k) are used onto or under graphene [17, 26–29], configurations with suspended graphene [23, 28, 30–32] or substrate-less graphene, or at encapsulating (embedding) graphene in dielectric materials, such as boron nitride, with lattice matched [28, 33–35], have maximal mobility. When graphene is embedded in dielectric materials, the strong Coulomb scattering increases the electrical mobility [28]. In addition, suspended graphene eliminates substrate-induced carrier scattering. Hybrid structures based on the use of an ultrathin benzocyclobutane (BCB) polymer as a buffer layer to reinforce the top gate of graphene used in field-effect transistors [32].

GFETs [26, 27, 35] and negative differential resistance (NDR) devices [36–41] exploit the previously described outstanding physical properties. These devices can work in the sub-millimetre and terahertz region [8]. Four different configurations to implement GFETs were proposed, such as back-gate GFETs, top-gate GFETs, wrap-around GFETs and suspended GFETs to design electronic devices; unfortunately, wrap-around GFETs still have no real implementation [23]. The readers are suggested to read the previous work for more details about the GFETs. Flexible graphene field-effect transistors (GFETs) are being fabricated with graphene channels fully encapsulated in hexagonal boron nitride through a self-aligned fabrication scheme [35]. These devices present an outstanding DC and RF performance with high mechanical flexibility. Despite high mobility of the electrical carriers in graphene, the ambipolar conduction and quantum capacitance are the fundamental limitations of the graphene itself in the development of electronic devices [34]. In addition, device transconductance (g_m) and output conductance (g_{ds}) characteristics in GFETs until now have not achieved the performance of the silicon CMOS devices. Field-effect transistors (FETs) based on 2D-material-based heterostructures with MoS_2 channels, hBN as dielectric and graphene gate electrodes are being designed for logic circuits offering an adequate mobility and low power consumption, thereby replacing conventional materials such as silicon [42].

Negative differential resistance (NDR) is the essential mechanism of three-terminal electronic devices such as high-frequency oscillators, frequency multipliers, memories, quantum dots and fast switches [36, 37, 40]. These devices offer better properties that conventional two-terminal NDR devices such as independence of quantum tunnelling and the gate electrode can be used to control the current density and the output power of the AC oscillation [37]. Moreover, tunnel diodes and tunnelling FETs can be developed using graphene with the effect of negative differential resistance to design high-performance devices for either analogue or digital applications [39]. These devices exploit the peak current and the peak-to-valley ratio which are strongly enhanced and weakly sensitive to the length fluctuations of the transition region, owing to the graphene working as the active material. Moreover, vertical transistors based on multilayers of graphene can be developed for applications in logic circuits, high-speed electronics and as barristors [38]. Circuits based on GFETs exploiting the effects of negative differential resistance (NDR) at room temperature without any technological doping step can be integrated with silicon-based circuits in the same process [41]. These circuits can be applied for developing amplifiers, oscillators, memories, switches, etc.

An interesting technological alternative is the use of three-dimensional printing of graphene scaffolds for electronic applications from a liquid ink [43]. These structures make use of composite materials based on polymers and graphene, which have potential applications in wearable and implantable nanoelectronics, and in devices such as transistors, supercapacitors, transparent conductors, interconnects and gas sensors. Mathematical modelling is being used to predict the electrical behaviour of the graphene that will be used in the design of electronics devices [44–47]. Increasing the width of graphene nanoribbons used in field-effect transistors results in an increment in the leakage current and subthreshold swing and decrease in their I_{ON}/I_{OFF} ratio [44]. It is possible to increase the I_{ON}/I_{OFF} ratio and subthreshold swing in graphene nanoribbon field-effect transistors using single-vacancy defects [46]. These defects increase the band gap of the graphene, as is demonstrated by theoretical studies using computer simulation. Dual-gate graphene nanoribbon field-effect transistor (DG-GNRFET) under local uniaxial strain in source and drain regions as a device suitable for switching applications [45]. Models based on 2D Poisson atomistic mode-space approach and Schrödinger equations within the Non-Equilibrium Green's (NEGF) are used to predict a high on-current and on-off ratio which is necessary for digital integrated circuits. An exhaustive study of the mathematical expressions of the electrical parameters of devices based on graphene is achieved using computer simulation with the aim of knowing the importance of this tool to predict the behaviour of field-effect transistors based on graphene (GFETs) [47]. In this work, a frequency analysis is realized to know the cut-off frequency (f_T) and maximum frequency (f_{max}) of the RF field-effect transistors based on graphene using different mathematical models. Moreover, the negative differential resistance (NRD) effect presented in GFET is completely analysed in the same work.

Interconnects refer to the physical connecting medium between several electrical nodes in a semiconducting chip to transmit signals from one point to another without any distortion [5, 24]. Depending on the orientation of carbon atoms on the edge of the graphene sheet, graphene nanoribbons (GNRs) can be either armchair or zigzag. Zigzag GNR always has metallic behaviour, whereas armchair GNR can have either semiconducting or metallic depending on geometry (chirality). An illustrative schematic of the different types of graphene nanoribbons is presented by the author in [47] for the reader, where pathways of electrical conductivity are better understood. Several layers of interconnects are required between devices; these can be horizontal and/or vertical [24, 48]. A vertical interconnection is called via; it is used to make connections between different horizontal levels in an integrated circuit to connect device to device, device to system or system to system [23]. For interconnecting applications, zigzag GNR is proposed for the future generation of VLSI circuits, due to its metallic property [5, 48].

3. Applications in Analogue Radiofrequency (RF) Devices and Integrated Circuit Design

Graphene offers the better prospects for developing flexible transistors based on 2D atomic sheets with good electrical and mechanical properties to implement electronic devices such as analogue RF devices, with a performance similar to that of the Si-CMOS technology, but on arbitrary plastic substrates [17]. Graphene nanoribbons with reduced width exhibit a low electrical mobility and high electron energy levels which increase gate leakage current and the large contact resistance between them and the metal contacts. Thus, graphene is not an ideal candidate for digital applications; but graphene is suitable for radiofrequency because RF transistors do not necessarily need to be turned completely off [16]. RF devices based on graphene have received much attention due to the significant progress that has been achieved in the last decade to implement wafer-scale-integrated amplifier circuits with voltage amplification until 20 dB with field-effect transistors operating with an intrinsic cut-off frequency above 300 GHz [49]. Graphene-based RF field-effect transistors (FETs) can be used to implement RF circuits with both cut-off frequencies f_T and maximum oscillation frequencies f_{max} working slightly above a few GHz [50]. Graphene has the potential to offer third-order linearity, at least, comparable to carbon nanotube-based field-effect transistors (CNFETs) and metal-oxide semiconductor field-effect transistors (MOSFETs), but it, unfortunately, suffers from worse second-order linearity. In addition, its load-resistance dependency is intimately tied to the lack of a band gap and linear density of states (DOS) of graphene [51]. Strategies such as increasing the graphene quality lead to increasing mobility, reduce contact resistance, and a good electrostatic control of the channel, and therefore, its drain-source current (I_{DS}) and transconductance (g_m) of the field-effect transistor [52].

The set of analogue RF devices and circuits, where graphene can be used, includes a very wide variety of RF ICs, where the entire RF signal chain is covered from DC to beyond hundreds of GHz [53]. The use of the ambipolar transport properties and high carrier mobility of graphene are exploited to design nonlinear electronics for RF applications including high-speed transmitters and receivers in a sensor network, satellite communications and radar systems [54]. Moreover, graphene has a great potential in RF communication electronics in the development of low noise amplifiers, frequency multipliers and resonators [53]. Some applications are mixers of microwaves and millimetre waves [54, 55], wafer-scale integrated graphene amplifier circuit [49], filters, absorbers and antennas with high-impedance surface [56].

Different mixers have been developed based on GFETs operating in the range of MHz [49, 53–55]. Due to the symmetrical ambipolar conduction of the graphene, graphene-based mixers can effectively suppress odd-order intermodulation and lead to lower spurious emissions in the circuit [54]. Graphene offers competitive advantages in RF mixers such as high conversion loss (CL) over the frequency range in GHz, good current on-off ratio, narrow bandwidth and better linearity. A mixer was designed based on microstrip technology using an array of bow-tie-structured graphene with performance better than those fabricated with other technologies.

Graphene top-gate transistors can be used as amplifiers to generate signal amplification [6]. Graphene voltage amplifiers present better high-gain signal amplification on conventional loads at room temperature in a frequency range surpassing classical values of their technological predecessors. Even frequency multipliers based on graphene can operate at 1.4 GHz [6].

Graphene-based two-dimensional laky-wave antenna (LWA) allows both frequency tuning and beam steering in the terahertz band [56]. These antennas can be used in the development of smart systems such as tunable transceivers and sensors because of its high directivity and frequency reconfiguration. Radar applications are possible, as the operating frequencies are > 100 GHz [11], where synthesis method has a direct effect on maximum frequency achieved by electronic devices. Graphene plasmons, quanta of the collective charge-density waves excited by two-dimensional carriers in graphene, can dramatically increase the light (THz photons) and matter (graphene) interaction, leading to "giant THz gain" [25].

In particular, polymer composites containing graphene are being studied by the author to be used as electromagnetic interference (EMI) shielding due to their unique combination of electrical conduction, polymeric flexibility and lightweight [57]. These materials exhibit moderately high electrical conductivity and low permittivity. The aspect ratio, orientation and the weight percentage of graphene have a direct effect on electromagnetic interference shielding of the resultant composite. These electromagnetic waves are not desired as they modify the electrical and magnetic behaviour of the electronic devices.

4. Applications in electrochemical energy systems and photonics

Graphene and its derivatives can be used in electrochemical energy systems requiring conversion and storage function such as batteries, fuel cells, and supercapacitors [58]. Numerous studies have been conducted to describe the advances achieved by researchers in energy applications using graphene as an active material [59, 60]. Mechanical properties such as mechanical resistance and flexibility can be exploited to design bendable, foldable and/or stretchable devices for flexible energy conversion and storage. The main applications of the graphene are photovoltaic devices (solar cells) [60], fuel cells [61], nanogenerators [62], supercapacitors [59] and batteries [58–60, 63]. These devices are potentially applied in roll-up displays, electronic papers, touch screens, active radiofrequency identification tags, wearable sensors and implantable medical devices, which form part of the applications of wearable and portable electronics. In addition, those materials used in these applications should offer high electrical and/or ionic conductivities, large specific surface areas and excellent chemical, photochemical and/or electrochemical stabilities. Graphene, graphene derivatives, and their composites fulfil these requirements, and now they are being used to design novel electronic devices for energy applications [64].

Graphene-based materials are used as transparent conductive electrodes or electron acceptors in solar cells, or as current collectors, electrodes, active materials or conductive electrodes for energy storage devices [60]. Advantageous properties of the graphene with respect to conventional materials such as metals and ceramics are useful for these applications, such as lighter weight densities, adequate flexibility, better optical transparency, higher optical, chemical and/or electrochemical stabilities, larger specific surface areas and higher electrical conductivity [8]. A thin film of graphene is semitransparent to the visible and NIR regions, whereas thick films are opaque. The transmittance and electrical conductivity of the graphene can be tuned by varying the thickness of the films and the degree of chemical reduction [28]. An ideal sheet of graphene exhibits sheet resistance of 6kΩ/□″ with nearly constant optical transparency of 98% in the visible-IR range. Graphene compared with the indium tin oxide (ITO) films has high strength, flexibility and chemical stability, and its production is less expensive [14].

Graphene can be used as an active material in solar cells only in *n*- and *p*-type semiconducting behaviour where a band gap of 1.4 eV is used when solar energy is used for the illumination. Chlorine added on both sides of the surface of graphene generates a band gap of 1.2 eV, while hydrogen placed in the same way presents a band gap of 2.54 eV [22]. Photovoltaic (PV) devices require very demanding specifications such as optical transmittance ($T > 85\%$) and sheet resistance ($R < 50\ \Omega/\square$). Graphene has been proposed as an ideal material to replace transparently and conductive oxides such as zinc oxide (ZnO), indium-tin oxide (ITO) and tin dioxide (SnO_2). However, further studies must be carried out for fulfilling such technological requirements [59]. A Schottky junction solar cell with modified graphene films and silicon pillar arrays provide a conversion efficiency of up to 7.7%. Heterojunction solar cells based on graphene/semiconductor can achieve conversion efficiency up to 9.2% [22]. In the case of dye-sensitized solar cells (DSSC), a lot of work must be done to incorporate graphene: (1) the graphene surface must be functionalized without affecting its work function so that the active layer can be attached to the surface of the graphene layer, (2) since graphene is hydrophobic, it must be hydrophilic to be used in solution with organic dye or electrolytes, and (3) an ohmic-type contact must be created between active layer of the solar cell and the graphene in the electrodes [22]. Graphene has been used in electrodes of Schottky cells, CdTe cells, dye-sensitized cells, organic cells and hybrid solar cells [65]. Doping and tailoring of graphene are strategies very useful for tuning electronic structure and work function, which are the key approaches for graphene to be assembled into photovoltaic cells.

Graphene and its derivatives have a strong impact on the development of electrodes and electrode supports in energy storage devices, due to their high surface area, improved porosity, tunable electrical conductivity and high mechanical strength [58]. Fortunately, the development of graphene-based materials is in its infancy, and the actual deficiencies can be overcome with the aim of achieving better performance. Numerous techniques are being experimentally tested for fabricating precise nanostructures with defined dimensions, and self-assembly techniques allow improve their physicochcmical and electrochemical properties [64].

Lithium-ion batteries must have high energy density, high voltage, long cycle life, light weight, and good environmental stability [58]. Graphene is used as an anode, owing to its amenability for reversible intercalation/deintercalation process with metal ions and in particular, lithium ions. The functional groups on graphene make it highly electronegative, thus resulting in selective interaction with cationic species. Graphene presents large capacity, high rate capability and excellent cycling stability, which facilitate the access of electrolyte and rapid diffusion of Li^+ ions and electrons and these deliver a large reversible capacity [13]. Rechargeable lithium-sulfur (Li-S) batteries to be optimized in their performance, such as high energy density, require novel materials such as graphene. This material is being used in sulphur positive electrodes, lithium negative electrodes and as an interlayer [63]. In the case of cathodes, now rarely can deliver a discharging capacity under high current densities, which is theoretically valued as 1673 mAh/g. However, there exists the possibility of improving its performance at synthesizing graphene sheets with controlled compositions, sizes and structures that can be required to obtain high electrical conductivity and high specific surface area possible only theoretically. With the aim of optimizing the cycling stability and rate capability of the Li-S batteries, functionalized graphene-based interlayers can be used for intercalating lithium ions among electrodes in the battery [64]. In a similar way, graphene is being used as a medium to load sulphur into battery during long cycle life to offer high energy density with an average voltage of 3.5 V. When graphene interlayers are used in batteries, there exists the possibility of restacking of these layers; for alleviating this problem, solid nanoparticles of Si, CuO, Fe_2O_3, SnO_2, Co_3O_4 or Mn_3O_4 can be used [15]. Graphene anodes experience significant irreversible capacity losses during charge/discharge cycling, mainly due to the restacking of graphene layers.

Supercapacitors serve as portable energy sources with smaller size, more flexible packaging, lighter weight, longer life, higher power capability, wide thermal operating range and more efficiency that conventional lead-acid or alkaline battery [66]. They store electrical energy in a capacitive form and where electrochemical double layer capacitors are formed at the electrode-electrolyte interface [58]. Specific capacitance and performance characteristics of graphene-based capacitors depend mainly on the route employed for preparing electrode material. Graphene has been used extensively as a material for electrodes used in supercapacitors. Efficient supercapacitors or hydrogen storage materials can exploit graphene, thanks to its high specific surface area (SSA) with theoretical values of 5000 m^2/g (considering the incorporation of holes into graphene), although the best state of the art is of only 3000 m^2/g [59, 64]. Graphene achieves an ideal capacitance of 200–500 F/g which depends on the surface area, pore size (both previous qualities are improved by chemical activation treated with alkali) and the electrical conductivity of the material (chemical doping to increase the carrier concentration) [22].

Fuel cells convert continuously supplied fuel to electricity, and particularly graphene is used as a catalyst support material for oxidation/reduction reactions [58]. Graphene and its derivatives have been considered as one of the most promising alternatives as electrode materials in energy-related devices, since they allow the oxidation of hydrogen and hydrogen-containing gases (e.g., methanol, ethanol, etc.) and/or the reduction of oxygen and oxygen-containing gases (such as air) in fuel cells [66]. Nitrogen-doped graphene has a good electrocatalytic activity for oxygen reduction and graphene loaded with Fe or Co enhances the electrocatalytic activity of the fuel cells [22]. This electrical activity involves the electron transfer out of or into the graphene sheets from the surrounding environment, due to the high electrical conductivity, large SSA, profuse interlayer structure and abounding functional groups involved [66]. Graphene is used as catalyst supports since it maximizes the availability of nanosized electrocatalyst surface area for electron transfer but also provide better mass transport of reactants to the electrocatalysts [65]. In addition, it facilitates efficient collection and transfer of electrons to the collecting electrode surface. The solubility of graphene oxide in different solvents allows it to be uniformly deposited onto a wide range of substrates in the form of thin films [61].

Graphene electrodes present high carrier mobility, which leads to high on/off ratio of the output current of graphene-based nanogenerators. Graphene provides electrical and structural stability under external mechanical loads such as bending and rolling. Graphene-based room-temperature (RT) nanogenerators can be used to develop self-powered RT device applications such as flexible self-powered touch sensors, wearable artificial skins, fully rollable display mobile devices and battery supplements for wearable cellular phones [62].

Graphene and its derivatives owing to their electronic and optical properties are ideal options for photonic and optoelectronic applications [67, 68]. The optical transparency and electrical conductivity of graphene can be exploited for many photonic devices [58]. Flexible and transparent optoelectronic devices based on graphene are transparent displays, solar cells and wearable electronics [21]. To ensure a good performance of these devices, it is necessary to integrate diverse classes of 2D materials, for example, graphene, with distinct physical properties. Graphene shows photonic properties such as absorption of a significant fraction of incident white light, strong tunable interband transitions and high contrast ratio [68]. In addition, it has a low broadband absorption which is ideal to transparent conductors [21]. Among different applications in these areas are transparent electrodes, touch screens, organic light emitting diodes (OLEDs), etc. Graphene-based transparent electrodes can be developed on flexible substrates for solar cells and the previously mentioned applications. In addition, graphene electrodes can be used in organic field-effect transistors (OFETs), resistive switching devices and molecular junction devices, thanks to the favourable interfacial contact between organic materials and electrochemical functionalization with graphene [69]. Touch screens require graphene-woven fabric to develop smart self-sensing elements based on piezo resistors directly transferred onto flexible substrates such as poly(dimethylsiloxane) (PDMS) [70]. Organic light-emitting diodes (OLEDs) are benefiting significantly from graphene-based transparent conducting electrodes (TCEs) where thin films of semiconducting metal oxides such as MoO_3 or WO_3 cover graphene [71]. The oxide coating provides effective graphene doping, ideal alignment of the transport levels at the graphene interface, effective wetting and graphene protection during etching and patterning.

5. Applications in gas sensors

Intensive research interest based on nanotechnology, for developing gas sensors more sensitive, with fast response and better stability, is being driven [72]. Gas sensors are based on chemiresistors (two-terminal graphene devices) and FETs with 1D nanostructures (three-terminal transistor-like structures). Graphene can play an important role in the development of chemical sensors due to its excellent chemical and surface properties derived from their chemical composition and the high-aspect ratio between its length and width [19]. Owing to these nanomaterials, it is possible to detect parts per billion or parts per trillion in comparison with their technological predecessors which could detect only part per million. With the aim of achieving it, defects or imperfections must be introduced to the sp^2 configuration of graphene to be used in the design of chemiresistors and chemical field-effect transistors (chemFETs) [22]. Gas sensors based on pristine graphene are less sensitive to analyte molecules because adsorbate binds to point defects, which have low resistance pathways around them [73]. Therefore, the conductance of graphene is more sensitive to the geometry and types of defects rather than their concentration. In addition, graphene must be cut into ribbons of width comparable to the line defect dimensions to offer superior performance as gas sensors. Graphene materials due to its superior properties such as thermoelectric conduction, surface area and mechanical strength, have inspired huge interest in sensing of various chemical species [74]. In addition, graphene can be modified to achieve high sensitivity and provide good selectivity for particular gases through methods such as using dopants and defects, decoration with metal/metal oxide nanoparticles and functionalization with polymers. The adsorption of a transition metal on graphene is one of the most studied due to the variety of promising materials for gas sensors [72]. Graphene's surface can be modified to lead to functional activity to increase the detection limit and response time at ambient temperatures, which are key parameters for an enhanced gas sensor. Unfortunately, the large-scale production of graphene-based gas sensors with high and uniform quality continues being a challenge in the electronic industry. Furthermore, novel strategies not explored to date must be exploited so that newer dopants, functional molecules and fabrication methods can be introduced.

Different graphene-based hybrids can be used in the development of chemiresistive gas sensors such as graphene with noble metals (such as platinum (Pt), palladium (Pd) and silver (Ag)), graphene with 3D, 2D, 1D or 0D metal oxides, graphene with conducting polymers (such as polythiophene (PTh), polyaniline (PANI) and polypyrrole (PPy)) and ternary graphene-based hybrids (where noble metal-metal oxide, noble metal-conducting polymers or metal oxide-conducting polymers are hybridized with graphene to jointly exploit their advantages) [75]. Metal oxides such as SnO_2, ZnO, WO_3, Cu_2O and Co_3O_4 are being used in hybrid materials based on graphene to develop toxic gas sensors for analytes such as CO, NO_x and NH_3 [76]. Some difficulties for its implementation on a large scale are a lack of reproducibility, nonuniform thickness of the graphene, high sheet resistance and relative inertness to hydrophilic atmospheres.

Water acts as an electron acceptor when adsorbed on the graphene surface which is accomplished by hole injection [28]. A *p*-type doping is used when the electron affinity of the adsorbate molecule (water) is greater than the work function of the substrate (graphene).

6. Prospects of graphene in electronics

Graphene is the cornerstone that experts in science and engineering materials have to implement innovative electronic devices and applications. A key to success in such applications is the development of novel methods to produce large quantities of graphene with high repeatability and quality. Researchers around the world are looking for alternative technological solutions for electronic devices to achieve maximum efficiency of all physicochemical properties that have the graphene. The modification of the bandgap is one of the main strategies to promote the extensive use of graphene. This advancement will allow a much wider range of applications not developed so far are reached, and where semiconductor materials have been exploited tremendously. The scalability of the production and processing convenience are important precursors to convert graphene and other two-dimensional materials (2D) in the material par excellence for the development of electronic devices in the XXI century. Aspects such as the control of the thickness of graphene, unusual rotational graphene stacking, and the relationship between the structure and electronic properties between graphene and its substrate must be clearly understood. Mathematical modelling of electronic devices based on graphene and its derivatives should be extended in order to broaden the understanding of the effects and physicochemical properties of the interaction between graphene and different materials such as metals, ceramics and polymers to produce hybrid materials, composite materials and functional materials, which have direct application in the development of innovative electronic devices.

7. Conclusions

A few decades ago, the potential of the electronics industry depended entirely on silicon; new materials have now been introduced to increase efficiency, capacity and speed of information processing in the electronics industry such as carbon allotropes such as carbon nanotubes and graphene. Actually, in electronics, graphene is used in the manufacture of supercapacitors, batteries, field-effect transistors, solar cells, light-emitting diodes, transparent, covered electrodes for electrostatic dissipation and/or electromagnetic interference shielding. Graphene's potential has not been fully associated with the development of materials science and engineering. The use of graphene in the electronic industry will be extended in the design of new electronic devices being applied either individually or as a component within a composite, hybrid or functional material. The cointegration of graphene and semiconducting

2D materials forming composite, hybrid or functional materials on the same flexible substrate will fulfil all electrical properties required by materials used in electronic industry at the thin-film limit. Although substantial progress has already been achieved to lead graphene to practical applications in electronic industry, however, a lot of work must be realized to consolidate the position of the graphene as the electronic material of this century. More studies for tuning electrical properties of the graphene and its derivatives (composites, hybrid, hierarchical or functional materials) could lead to the first large-scale applications based on graphene.

Acknowledgements

The author acknowledges funding from the CONACYT (contract no. 152524, basic science), Tecnológico Nacional de México (contract no. 284.15-PD), and Instituto Tecnológico Superior de Irapuato (ITESI).

Author details

Rafael Vargas-Bernal

Address all correspondence to: ravargas@itesi.edu.mx

Materials Engineering Department, Higher Technological Institute of Irapuato, Irapuato, Guanajuato, México

References

[1] Fiori G., Bonaccorso F., Iannaccone G., Palacios T., Neumaier D., Seabaugh A., Banerjee S.K., Colombo L. Electronics based on Two-Dimensional Materials. Nature Nanotechnology. 2014; 9(10):768–778. DOI:10.1038/nnano.2014.207

[2] Yuan W., Chen J., Shi G. Nanoporous Graphene Materials. Materials Today. 2014; 17(2): 77–85. DOI:10.1016/j.mattod.2014.01.021

[3] Chang H., Wu H. Graphene-based Nanomaterials: Synthesis, Properties, and Optical and Optoelectronic Applications. Advanced Functional Materials. 2013; 23:1984–1997. DOI:10.1002/adfm.201202460

[4] Ferrari A.C., Bonaccorso F., Fal'ko V., Novoselov K.S., Roche S., Bøggild P., Borini S., Koppens F.H.L., Palermo V., Pugno N., Garrido J.A., Sordan R., Bianco A., Ballerini L., Prato M., Lidorikis E., Kivioja J., Marinelli C., Ryhänen T., Morpurgo A., Coleman J.N., Nicolasi V., Colombo L., Fert A., Garcia-Hernandez M., Bachtold A., Schneider G.F.,

Guinea F., Dekker C., Barbone M., Sun Z., Galiotis C., Grigorenko A.N., Konstantatos G., Kis A., Katnelson M., Vandersypen L., Loiseau A., Morandi V., Neumaier D., Treossi E., Pellegrini V., Polini M., Tredicucci A., Williams G.M., Hong B.H., Ahn J.-H., Kim J.M., Zirath H., van Wees B.J., van der Zant H., Occhipinti L., Di Matteo A., Kinloch I.A., Seyller T., Quesnel E., Feng X., Teo K., Rupesinghe N., Hakonen P., Neil S.R.T., Tannock Q., Löfwander T., Kinaret J. Science and Technology Roadmap for Graphene, related Two-Dimensional Crystals, and Hybrid Systems. Nanoscale. 2015; 7(11):4587–5062. DOI:10.1039/c4nr01600a

[5] Das D., Rahaman, H. Carbon Nanotube and Graphene Nanoribbon Interconnects. Boca Raton, FL, United States of America: CRC Press; 2015, 196 p. DOI: 10.1007/978-3-662-45711-5

[6] Yusoff A.R.b.M. (Ed). Graphene Optoelectronics: Synthesis, Characterization, Properties, and Applications. Weinheim, Germany: Wiley-VCH; 2015. 290 p. DOI: 10.1002/9783527677788

[7] Akinwande D., Petrone N., Hone J. Two-Dimensional Flexible Nanoelectronics. Nature Communications. 2014; 5:5678. DOI:10.1038/ncomms6678

[8] Choi W., Lee J.-W. (Ed.). Graphene: Synthesis and Applications. Boca Raton, FL, United States of America: CRC Press; 2012. 373 p. DOI:10.1201/b11259

[9] Tanaka K., Iijima S. Carbon Nanotubes and Graphene. Second Edition. Waltham, MA, USA: Elsevier; 2014. 446 p. DOI:10.1016/B978-0-08-098232-8

[10] Pérez E.M., Martin N. π-π Interactions in Carbon Nanostructures. Chemical Society Reviews. 2015; 44(18):6425. DOI:10.1039/c5cs00578g

[11] Warner J.H., Schäffel F., Bachmatiuk A., Rümmeli M.H. (Ed.). Graphene: Fundamentals and Emergent Applications. Oxford, United Kingdom: Elsevier; 2013. 449 p. DOI: doi: 10.1016/B978-0-12-394593-8

[12] Wang X., Shi G. An Introduction to the Chemistry of Graphene. Physical Chemistry Chemical Physics. 2015; 17(43):28484. DOI:10.1039/c5cp05212b

[13] Tiwari A., Syväjärvi M. Graphene Materials: Fundamentals and Emerging Applications. Hoboken, New Jersey, United States of America: John Wiley & Sons; 2015. 424 p. DOI:10.1002/9781119131816

[14] Zheng Q., Kim J.-K. Graphene for Transparent Conductors: Synthesis, Properties and Applications. New York, United States of America: Springer; 2015. 231 p. DOI: 10.1007/978-1-4939-2769-2

[15] Gao W. (Ed). Graphene Oxide: Reduction Recipes, Spectroscopy, and Applications. New York, USA: Springer; 2015. 154 p. DOI: 10.1007/978-3-319-15500-5

[16] Murali R. (Ed.). Graphene Nanoelectronics: From Materials to Circuits. New York, NY, United States of America, Springer; 2012. 271 p. DOI:10.1007/978-1-4614-0548-1

[17] Raza H. (Ed.). Graphene Nanoelectronics: Metrology, Synthesis, Properties and Applications. Heidelberg, Germany: Springer; 2012, 611 p. DOI: 10.1007/978-3-642-22984-8

[18] Pati S.K., Enoki, T., Rao C.N.R. (Ed.). Graphene and Its Fascinating Attributes. Danvers, MA, United States of America: World Scientific Publishers; 2011. 287 p. DOI: 10.1142/9789814329361

[19] Wong H.-S. P., Akinwande, D. Carbon Nanotube and Graphene Device Physics. Cambridge, United Kingdom: Cambridge University Press; 2011. 263 p. DOI:1017/CB09780511778124

[20] Katsnelson M.I. Graphene: Carbon in Two Dimensions. Cambridge, United Kingdom: Cambridge University Press; 2012. 364 p. DOI:1017/CB09780521195409

[21] Shafraniuk, S. Graphene: Fundamentals, Devices, and Applications. Danvers, MA, United States of America: Pan Stanford Publishing; 2015. 634 p. DOI: 10.4032/9789814303316

[22] Sharon M. Graphene: An Introduction to the Fundamentals and Industrial Applications. Hoboken, New Jersey, United States of America: Wiley; 2015. 317 p. DOI: 10.1002/9781118842577

[23] Vargas-Bernal R., Herrera-Pérez G. Carbon Nanotube- and Graphene based Devices, Circuits and Sensors for VLSI Design. In: Tlelo-Cuautle E., Tan S.X.-D., editors. VLSI Design. Rijeka, Croatia: Intech; 2012. p. 41–66. DOI:10.5772/38743

[24] Vargas-Bernal, R. Performance Analysis of Interconnects based on Carbon Nanomaterials for AMS/RF IC Design. In: Fakhfakh M., Tlelo-Cuautle E., Fino, M.H., editors. Performance Optimization Techniques in Analog, Mixed-Signal, and Radio-Frequency Circuit Design. Hershey PA, USA: IGI Global; 2015. p. 336–363. DOI: 10.4018/978-1-4666-6627-6.ch014

[25] Matsumoto, K. (Ed.). Frontiers of Graphene and Carbon Nanotubes: Devices and Applications. Tokyo, Japan: Springer; 2015. 295 p. DOI:10.1007/978-4-431-55372-4

[26] Liao L., Bai J., Qu Y., Lin Y.-C., Li Y., Huang Y., Duan X. High-k Oxide Nanoribbons as Gate Dielectrics for High Mobility Top-Gated Graphene Transistors. PNAS. 2010; 107(15): 6711–6715. DOI:10.1073/pnas.0914117107

[27] Vicarelli L., Vitiello M.S., Coquillat D., Lombardo A., Ferrari A.C., Knap W., Polini M., Pellegrini V., Tredicucci A. Graphene Field-Effect Transistors as Room-Temperature Terahertz Detectors. Nature Materials. 2012; 11(10):865–871. DOI:10.1038/nmat3417

[28] Skákalová V., Kaiser A.B. (Ed.). Graphene: Properties, Preparation, Characterisation and Devices. Cambridge, United Kingdom: Woodhead Publishing; 2014. 401 p. DOI: 10.1533/9780857099334

[29] Jin Z., Ma P., Wang S., Peng S., Zhang D., Shi J., Niu J., Yu G., Wang X., Li M. Hydroxyl-Free Buffered Dielectric for Graphene Field-Effect Transistors. Carbon. 2015; 86: 264–271. DOI:10.1016/j.carbon.2015.01.030

[30] Adam S., Sarma S.D. Transport in Suspended Graphene. Solid State Communications. 2008; 146(9-10): 356–360. DOI:10.1016/j.ssc.2008.03.021

[31] Bolotin K.I., Sikes K.J., Jiang Z., Klima M., Fudenberg G., Hone J., Kim, P. Stormer H.L. Ultrahigh Electron Mobility in Suspended Graphene. Solid State Communications. 2008; 146(9-10): 351–355. DOI:10.1016/j.ssc.2008.02.024

[32] Zang Y., Zhang F., Huang D., Gao X., Di, C.-A., Zhu D. Flexible Suspended Gate Organic Thin-Film Transistors for Ultra-Sensitive Pressure Detection. Nature Communications. 2015; 6(3): 6269 DOI:10.1038/ncomms7269

[33] Mayorov A.S., Gorbachev R.V., Morozov S.V., Britnell L., Jalil R., Ponomarenko L.A., Blake P., Novoselov K.S., Watanabe K., Taniguchi T., Geim A.K. Micrometer-Scale Ballistic Transport in Encapsulated Graphene at Room Temperature. Nano Letters. 2011; 11(6): 2396–2399. DOI:10.1021/nl200758b

[34] Chari T., Meric I., Dean C., Shepard K. Properties of Self-Aligned Short-Channel Graphene Field-Effect Transistors based on Boron-Nitride-Dielectric Edge Contacts. IEEE Transactions on Electron Devices. 2015; 62(12):4322–4326. DOI:10.1109/TED.2015.2482823

[35] Petrone N., Chari T., Meric I., Wang L., Shepard K.L., Hone J. Flexible Graphene Field-Effect Transistors Encapsulated in Hexagonal Boron Nitride. ACS Nano. 2015; 9(9): 8953–8959. DOI:10.1021/acsnano.5b02816

[36] Ren H., Li Q.-X., Luo Y., Yang J. Graphene Nanoribbon as a Negative Differential Resistance Device. Applied Physics Letters. 2009; 94(17):173110. DOI:10.1063/1.3126451

[37] Wu Y., Farmer D.B., Zhu W., Han S.-J., Dimitrakopoulos C.D., Bol A.A., Avouris P., Lin Y.-M. Three-Terminal Graphene Negative Differential Resistance Devices. ACS Nano. 2012; 6(3): 2610–2616. DOI:10.1021/nn205106z

[38] Britnell L., Gorbachev R.V., Geim A.K., Ponomarenko L.A., Mishchenko A., Greenaway M.T., Fromhold T.M., Novoselov K.S., Eaves L. Resonant Tunnelling and Negative Differential Conductance in Graphene Transistors. Nature Communications. 2013; 4(4): 1794. DOI:10.1038/ncomms2817

[39] Nguyen V.H., Saint-Martin J., Querlioz D., Mazzamuto F., Bournel A., Niquet Y.-M., Dollfus P. Bandgap Nanoengineering of Graphene Tunnel Diodes and Tunnel Transistors to Control the Negative Differential Resistance. Journal of Computational Electronics. 2013; 12(2):85–93. DOI:10.1007/s10825-013-0434-2

[40] Jiang C., Wang X.-F., Zhai M.-X. Spin Negative Differential Resistance in Edge doped Zigzag Graphene Nanoribbons. Carbon. 2014; 68: 406–412. DOI:10.1016/j.carbon.2013.11.017

[41] Sharma P., Bernard L.S., Bazigos A., Magrez A., Ionescu A.M. Graphene Negative Differential Resistance Circuit with Voltage-Tunable High Performance at Room Temperature. IEEE Electron Device Letters. 2015; 36(8):865–867. DOI:10.1109/LED. 2015.2445858

[42] Lee G.-H., Yu Y.-J., Cui X., Petrone N., Lee C.-H., Choi M.S., Lee D.-Y., Lee C., Yoo W.J., Watanabe K., Taniguchi T., Nuckolls C., Kim P., Hone J. Flexible and Transparent MoS_2 Field-Effect Transistors on Hexagonal Boron Nitride-Graphene Heterostructures. ACS Nano. 2013; 7(9): 7931–7936. DOI:10.1021/nn402954e

[43] Jakus A.E., Secor E.B., Rutz A.L., Jordan S.W., Hersam M.C., Shah R.N. Three-Dimensional Printing of High-Content Graphene Scaffolds for Electronic and Biomedical Applications. ACS Nano. 2015; 9(4):4636–4648. DOI:10.1021/acsnano.5b01179

[44] Banadaki Y.M., Srivastava A. Investigation of the Width-dependent Static Characteristics of Graphene Nanoribbon Field Effect Transistors using Non-Parabolic Quantum-based Model. Solid-State Electronics. 2015; 111: 80–90. DOI:10.1016/j.sse.2015.05.003

[45] Moslemi M.M., Moravej-Farshi M.K., Sheikhi M.H. Improving Ion/IOFF in Dual-gate Graphene Nanoribbon Field_Effect Transistors using Local Uniaxial Tensile Strain. Physica E. 2015; 68: 143–148. DOI: 10.1016/j.physe.2014.12.032

[46] Nazari A., Faez R., Shamloo H. Improving I_{ON}/I_{OFF} and Sub-Threshold Swing in Graphene Nanoribbon Field-Effect Transistors using Single Vacancy Defects. Superlattices and Microstructures. 2015; 86: 483–492. DOI: 10.1016/j.spmi.2015.08.018

[47] Vargas-Bernal, R. Advances in Computational Modeling of Electronic Devices based on Graphene. IEEE Journal on Emerging and Selected Topics in Circuits and Systems. 2015; 5(1):109–116. DOI: 10.1109/JETCAS.2015.2398219

[48] Chen J., Bo Z., Lu G. Vertically-Oriented Graphene: PECVD Synthesis and Applications. Heidelberg, Germany: Springer; 2015. 121 p. DOI:10.1007/978-3-319-15302-5

[49] Wu Y., Jenkins, K.A., Valdes-Garcia A., Farmer D.B., Zhu Y., Bol A.A., Dimitrakopoulos C., Zhu, W., Xia F., Avouris P., Lin Y.-M. State-of-the-Art Graphene High-Frequency Electronics, Nano Letters. 2012; 12(6):3062–3067. DOI:10.1021/nl300904k

[50] Yeh C.-H., Lain Y.-W., Chiu Y.-C., Liao C.-H., Moyano D.R., Hsu S.S.H., Chiu P.-W. Gigahertz Flexible Graphene Transistors for Microwave Integrated Circuits. ACS Nano, 2014; 8(8):7663–7670. DOI: 10.1021/nn5036087

[51] Alam A.U., Holland K.D., Wong M., Ahmed S., Kienle D., Vaidyanathan M. RF Linearity Performance Potential of Short-Channel Graphene Field-Effect Transistors. IEEE Transactions on Microwave Theory and Techniques. 2015; 63(12): 3874–3887. DOI: 10.1109/TMTT.2015.2496295

[52] Frégonèse S., de Matos M., Mele D., Maneaux C., Happy H., Zimmer T. Source-Pull and Load-Pull Characterization of Graphene FET. Journal of the Electron Devices Society. 2015; 3(1):49–53. DOI: 10.1109/JEDS.2014.2360408

[53] Palacios T., Hsu A., Wang H. Applications of Graphene Devices in RF Communications. IEEE Communications Magazine. 2010; 48(6): 122–128. DOI: 10.1109/MCOM.2010.5473873

[54] Wang H., Hsu A., Wu J., Kong J., Palacios T. Graphene-based Ambipolar RF Mixers. IEEE Electron Device Letters. 2010; 31(9): 906–908. DOI: 10.1109/LED.2010.2052017

[55] Habibpour O., Vukusic J., Stake J. A 30 GHz Integrated Subharmonic Mixer based on a Multi-Channel Graphene FET. IEEE Transactions on Microwave Theory and Techniques. 2013; 61(2): 841–847. DOI:10.1109/TMTT.2012.2236434

[56] Wang X.-C., Zhao W.-S., Hu J., Yin W.-Y. Reconfigurable Terahertz Leaky-Wave Antenna using Graphene-based High-Impedance Surface. IEEE Transactions on Nanotechnology. 2015; 14(1): 62–69. DOI: 10.1109/TNANO.2014.2365205

[57] Vargas-Bernal, R. Performance Analysis of Electromagnetic Interference Shielding based on Carbon Nanomaterials used in AMS/RF IC Design. In: Fakhfakh M., Tlelo-Cuautle E., Fino, M.H., editors. Performance Optimization Techniques in Analog, Mixed-Signal, and Radio-Frequency Circuit Design. Hershey PA, USA: IGI Global; 2015. p. 268–294. DOI: 10.4018/978-1-4666-6627-6.ch011

[58] Rao C.N.R, Sood A.K. Graphene: Synthesis, Properties, and Phenomena. Weinheim, Germany: Wiley-VCH; 2013. 426 p. DOI: 10.1002/9783527651122

[59] Quesnel E., Roux F., Emieux F., Faucherand P., Kymakis E., Volonakis G., Giustino F., Martín-García B., Moreels I., Gürsel S.A. Graphene-based Technologies for Energy Applications, Challenges and Perspectives. 2D Materials. 2015; 2(3):030204. DOI: 10.1088/2053-1583/2/3/030204

[60] Wang X., Shi G. Flexible Graphene Devices related to Energy Conversion and Storage. Energy & Environmental Science. 2015; 8(3): 790–823. DOI: 10.1039/c4ee03685A

[61] Tiwari A., Balandin A.A. (Ed.). Innovative Graphene Technologies Vol. 2: Evaluation and Applications. Shroshire, United Kingdom: Smithers Rapra; 2013. 564 p.

[62] Choi D., Choi M.-Y., Choi W.M., Shin H.-J., Park H.-K., Seo J.-S., Park J., Yoon S.-M., Chae S.J., Lee Y.H., Kim S.-W., Choi J.-Y., Lee S.Y., Kim J.M. Fully Rollable Transparent Nanogenerators based on Graphene Electrodes. Advanced Materials. 2010; 22(19): 2187–2192. DOI: 10.1002/adma.200903815

[63] Yu M., Li R., Wu M., Shi G. Graphene Materials for Lithium-Sulfur Batteries. Energy Storage Materials. 2015; 1: 51–73. DOI:10.1016/j.ensam.2015.08.004

[64] Yusoff, A.R.b.M. Graphene-based Energy Devices. Weinheim, Germany: Wiley-VCH; 2015. 463 p. DOI:10.1002/9783527690312

[65] Liu Z.P., Zhou X.F. Graphene: Energy Storage and Conversion. Boca Raton, Florida, United States of America, CRC Press; 2015. 318 p. DOI: 10.1201/b17757

[66] D'Souza F., Kadish K.M., (Ed.). Handbook of Carbon Nano Materials Vol. 6 Graphene - Energy and Sensor Applications. Danvers, MA, USA: World Scientific Publishing; 2014. 262 p. DOI:10.1142/8979

[67] Bonaccorso F., Sun Z., Hasan T., Ferrari A.C. Graphene Photonics and Optoelectronics. Nature Photonics. 2010; 4(9): 611–622. DOI: 10.1038/nphoton.2010.186

[68] Yamashita S., Saito Y., Choi J.H. Carbon Nanotubes and Graphene for Photonic Applications. Philadelphia, PA, United States of America: Woodhead Publishing; 2013. 434 p. DOI:10.1533/9780857098627

[69] Jo G., Choe M., Lee S., Park W., Kahng Y.H., Lee T. The Application of Graphene as Electrodes in Electrical and Optical Devices. Nanotechnology. 2012; 23(11): 112001. DOI: 10.1088/0957-4484/23/11/112001

[70] Lee X., Yang T., Li X., Zhang R., Zhu M., Zhang H., Xie D., Wei J., Zhong M., Wang K., Wu D., Li Z., Zhu H. Flexible Graphene Woven Fabrics for Touch Sensing. Applied Physics Letters. 2013; 102(16): 163117. DOI: 10.1063/1.4803165

[71] Kidambi P.R., Weijttens C., Robertson J., Hofmann S. Meyer J. Multifunctional Oxides for Integrated Manufacturing of Efficient Graphene Electrodes for Organic Electronics. Applied Physics Letters. 2015; 106(6): 063304. DOI: 10.1063/1.4908292

[72] D'Souza F., Kadish K.M., (Ed.). Handbook of Carbon Nano Materials Vol. 5 Graphene - Fundamental Properties. Danvers, MA, USA: World Scientific Publishing; 2014. 287 p. DOI:10.1142/8979

[73] Salehi-Khojin A., Estrada D., Lin K.Y., Bae M.-H., Xiong F., Pop E., Masel R.I. Polycrystalline Graphene Ribbons as Chemiresistors. Advanced Materials. 2012; 24(1): 53–57. DOI: 10.1002/adma.201102663

[74] Varghese S.S., Lonkar S., Singh K.K., Swaminathan S., Abdala A. Recent Advances in Graphene-based Gas Sensors. 2015; 218: 160–183. DOI: 10.1016/j.snb.2015.04.062

[75] Meng F.-L., Guo, Z., Huang X.-J. Graphene-based Hybrids for Chemiresistive Gas Sensors. Trends in Analytical Chemistry. 2015; 68: 37–47. DOI: 10.1016/j.trac. 2015.02.008

[76] Chatterjee S.G., Chatterjee S., Ray A.K., Chakraborty A.K. Graphene-Metal Oxide Nanohybrids for Toxic Gas Sensor: A Review. Sensors and Actuators B: Chemical. 2015; 221: 1170–1181. DOI: 10.1016/j.snb.2015.07.070

Chapter 2

Aspects of Nanoelectronics in Materials Development

Gaurav Pandey, Deepak Rawtani and
Yadvendra Kumar Agrawal

Additional information is available at the end of the chapter

http://dx.doi.org/10.5772/64414

Abstract

Nanotechnology is an enabling technology that potentially impacts all aspects of the chip-making practice from materials to devices, to circuits, and to system-level architecture. Nanoelectronics is an interdisciplinary division which refers to the use of nanotechnology in electronic components. The materials and devices used in nanoelectronics are so small that the interatomic interactions and quantum mechanical properties of such materials need to be studied extensively. Various electronic devices manufactured at nanoscale have been established: devices having negative differential resistance, switches which can be electrically configured, tunneling junctions, carbon nanotube (CNT) transistor, and unimolecular transistor. Some devices have also been linked together to form circuits proficient of performing functions such as logic functions and basic memory. Some of the widely used materials in nanoelectronics include zero-dimensional materials like quantum dots; one-dimensional materials like nanotubes and nanowires; nanoclusters and nanocomposites; carbon-based materials like carbon nanotubes (CNTs), fullerenes and graphene; etc. Plastic C nanoelectronics is also a prominent research area with collaboration between the materials science, chemistry, physics, nanotechnology, and engineering communities. As one of the most promising contenders, C nanostructures, either 2D graphene or quasi-1D CNTs, have unlocked entirely new standpoints concerning the C-based electronics. This chapter focuses on the approaches of nanotechnology toward nanoelectronics, materials used in nanoelectronics and the applications of nanoelectronics related to carbon-based materials in the field of thin-film transistors, printed electronics (PE), artificial skin and muscle, wearable electronics, flexible gas sensors, multifunctional and responsive elastomers, and plastic solar panels.

Keywords: Nanoelectronics, Semiconductors, Printed electronics, Electronic textile, Thin-film transistors

1. Introduction

In the twenty-first century, nanotechnology is evolving as cutting-edge technology and has implausible applications in physics, chemistry, biology, materials science, and medicine. The main push has been toward fabricating novel materials and investigating their properties by alteration in the particle size, shape, and distribution. This technology is an aiding technology which potentially influences nearly all aspects of the chip preparation from various kinds of materials to different devices, system-level architecture, and circuits. The technology is still in its initial phases, particularly in the field of conventional integrated circuit (IC) applications: logic computing and data storage. Examples of technologies which are engaged, presumably under molecular or quantum operations, may develop in the next 10–20 years [1].

There has been a firm growth of interest in nanoelectronics all through the past decade. It is an interdisciplinary division which talks about the use of nanotechnology in electronic components. It targets to enable the use of new approaches and materials to build electronic devices with feature sizes at the nanoscale level. The materials and devices used in nanoelectronics are so small that the interatomic contacts and quantum mechanical properties of such materials need to be considered comprehensively. Even though much of this work has a long-term explanation in the search for novel devices, most of the results are having scientific significance instead of engineering. Nanoelectronics has been considered most in the III–V semiconductors and principally in the Ga(Al)As alloy system where the construction and growth technologies are most established. Additionally, the transport properties of GaAs and the accessibility of very high-quality heterojunctions boost some potentially utilizable phenomena in a dimensional range available to high-performance lithography [2].

According to Moore's law, the computing power or the microprocessor speed has doubled up every complementary metal-oxide semiconductor (CMOS) generation or every 18 months, based on the sequential trend. This drift, which was witnessed first by examination of the past data, soon became a standard and objective for the growth of semiconductor technology and manufacturing. Yet, we have not touched the limit of computing power that is probable in the long run. Areas like advanced chip technology are concentrating on the development of nanoelectronic devices which are focused at discovering the limits to miniaturization of conventional transistors and revealing the phenomena which might lead to novel generations of devices operating on alternative principles. These miniaturized devices, beyond the Moore's law, are anticipated to be dependent on new, resourceful methods to implement spatially controlled and extremely functional nanoscale components manufactured by economical chemistry. The driving source in the last 3–4 eras of enhanced R&D efforts for advanced chip technology is the demand for computing power, along with that for data storage capacity, occasioning in improved performance and developed component packing density [3].

Nanoelectronics based on carbonaceous materials (plastic carbon nanoelectronics) is also a prominent research field with collaboration between various engineering communities and physics, chemistry, materials science, and nanotechnology. Since the advancement of plastic/organic electronics, there is a remarkable attention on the properties of the materials (based on carbon) owing to the considerable delocalization of π-electron in fullerenes and carbon nanotubes (CNTs). As one of the most promising contenders, C nanostructures, either 2D graphene [4,5] or quasi-1D CNTs [6], have unlocked entirely new standpoints concerning the C-based electronics. Through recent scientific advances in the field of nanoplastics which is a blend of the traditional plastics with emerging field of nanotechnology—many exhilarating results have established that polymers filled with carbon-based materials have exceptional properties including improved UV absorption, durability and strength, electrical conductivity, flame resistance, and reduced permeability over the last decade.

Many nanoscale electronic devices have been developed till date by innumerable researchers like tunneling junctions [7–9], devices using differential resistance (negative) [10], switches which can be electrically configured [11,12], CNT transistors [13,14], and unimolecular transistors [15,16]. Some devices have also been linked together to form circuits which are capable of carrying out functions like basic memory [11,12,17] and logic functions [18–21]. Nanowire lattices with ultrahigh density and circuits with semiconductor and metal nanowires have also been documented by Heath et al. The technique was based on transforming thin-film growth thickness control into planar wire arrays [22]. Architecture of computer which is grounded on nanoelectronics (also known as nanocomputers) has also been considered [23,24], though very restricted.

This chapter focuses on the approaches of nanotechnology toward nanoelectronics, materials used in nanoelectronics and the applications of nanoelectronics related to carbon-based materials in the field of thin-film transistors, printed electronics (PE), artificial skin and muscle, wearable electronics, flexible gas sensors, multifunctional and responsive elastomers, and plastic solar panels.

2. Nanotechnology approaches for nanoelectronics

There are mainly two types of approaches for the synthesis of nanomaterials used in nanoelectronics:

1. Top-down approach

2. Bottom-up approach

Top-down approach discusses about the slicing or successive cutting of a bulk material in order to get nano-dimensional particles. Bottom-up approach refers to the stockpile of a material from the bottom: atom by atom, molecule by molecule, or cluster by cluster. Attrition or milling is a characteristic top-down method in the production of nanoparticles, while the colloidal dispersion is a good illustration of bottom-up approach in the synthesis of nanoparticles. The major difficulty with top-down approach is the defectiveness of surface structure and significant crystallographic impairment of the processed patterns. These deficiencies lead to extra challenges in the device design and construction. But this approach leads to the bulk fabrication of nanomaterial. Regardless of the flaws produced by top-down approach, they will continue to play an imperative role in the production of nanostructures. When structures plunge into a nanometer scale, there is a minute chance for top-down approach. All the tools we have possessed are too large to deal with such miniature subjects. Bottom-up approach also assures a better chance to attain nanostructures with less defects and more consistent chemical composition. On the contrary, top-down approach presumably introduces internal stress, additionally to surface defects and contaminations (see **Figure 1**).

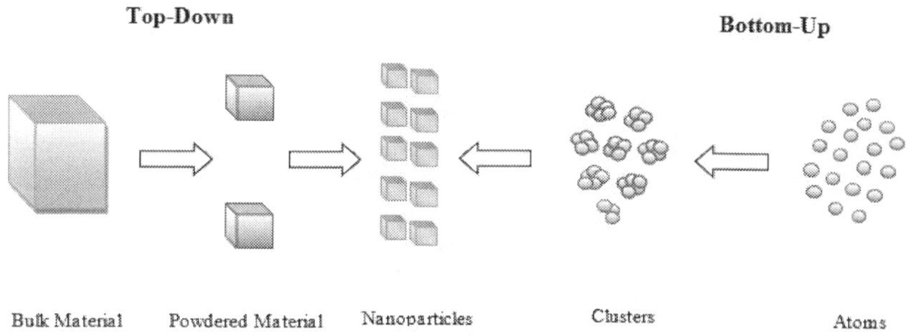

Figure 1. Approaches for the synthesis of nanomaterials.

The performance of Very Large Scale Integration circuits which are basically based on silicon has been progressively enhanced over the past few decades by scaling down the dimensions of the device, plus a virtually exponential development of microelectronics proficiencies is attained. Yet, maintaining this top-down trend for miniaturization is becoming very hard due to the basic technological and physical restrictions as well as the financial limitation. By contrast, using organic molecules as a building block for nanoscale devices has engrossed much consideration because the specifically controlled nanostructures may be formed economically by utilizing self-assembly of molecules. This bottom-up technology can possibly overcome the integral problem of the present silicon top-down technology [25].

Integrated circuit (IC) technology also utilizes the top-down methodology in which the interconnects and devices are prepared by procedures such as thin-film deposition, etching, photolithography, and metallization. Although the fabrication tool competencies keep refining with each technology generation, the complications rise histrionically as the dimensions of the components approach nanometer scale in which various physical effects would stop the tools from operating properly. These restrictions set the ultimate scaling barrier of silicon CMOS, irrespective of the great technological impetus built up over the last three decades. Bottom-up methodology employs chemistry which is not so expensive, to stimulate self-assembly of multifarious mesoscopic structures. There are some fundamental advantages related to bottom-up approach in nanotechnology, like extremely low-power dissipation, rich functionality, and high component packing density, which would fall perfectly on the trend of technological demand but unattainable by the top-down approach [1].

The fundamental differences between the traditional top-down semiconductor nanoscale technology and futuristic bottom-up molecular nanotechnology are listed in **Table 1**.

http://dx.doi.org/10.5772/64414

Top-down approach for semiconductor nanoscale technology	Bottom-up approach for molecular nanotechnology
Approach is from larger (bulk wafer) to smaller (chip)	Approach is from small (self-assembled structure) to large (chip)
Methods employed are pattern and etch	Methods employed are synthesis and self-assembly
Production cost is high	Production cost is low
Accessibility for material/device is less	Accessibility is more
Selection of material is stubborn	Selection of material is malleable
Design tool/infrastructure is developed	Design tool/infrastructure is open

Table 1. Comparison between two approaches in nanoelectronics.

Overall, considering the advantages of bottom-up approach over the top-down approach, it can be said that bottom-up technology is better than the top-down technology for the fabrication of various materials used in nanoelectronics.

3. Materials used in nanoelectronics

The nanoscale components used in nanoelectronics fall into two major categories:

1. Inorganic nanocrystals

2. Organic molecular components

1. Inorganic nanocrystals such as nanotubes and nanowires, named after their physical shape determined through the use of growth kinetics, are typically low-dimensional (up to nanoscale) structures. Substantial progress has been made on carbon nanotube (CNT) from the time of its discovery in the early 1990s [26]. Yet, hindrances such as difficulty in separating nanotubes, having different chiralities and properties, and process compatibility inhibit CNT from being readily used for chip applications, though advancement has been made in display [27], and sensing and thermal conducting applications. Li et al. [28] have developed a nanoelectrode assay grounded on vertically aligned multi-walled carbon nanotubes (MWCNTs) rooted in SiO_2 which is used for ultrasensitive DNA recognition. The use of aligned MWCNTs delivered a new bottom-up arrangement for fabricating trustworthy nanoelectrode arrays. Li et al. [29] have fabricated a gas sensor by the simple molding of single-walled carbon nanotubes (SWCNTs) on an interdigitated electrode (IDE) and offered for gas and organic vapor recognition at room temperature. The electrical response indicated well-defined and reproducible linear performance and a detection limit of <44 ppb for NO_2 and 262 ppb for nitrotoluene. Thermal contact conductance properties of CNF-Cu composite materials have also been established by Li and coworkers [30]. The composite material CNF-Cu had been verified at a variety of powers and pressures, yielding reasonable trends specifying the viability of their usage as a thermal interface material in both IC packaging and equipment-cooling applications. 1D semiconductor nanowires may find a number of uses in chip technology due to

exceptional compatibility of the process with the facilities of silicon fabrication, deterministic electrical properties, and variability of material selections.

2. Molecular components (organic) like molecular wires, single molecule, molecular monolayers, and supramolecules with different schemes are the contenders from other classes. Organic molecular components are prepared by amalgamation, and therefore less variation is probable in chemical composition and structural parameters. There are a large number of molecular species prevailing, possibly personalized for various device applications using surface molecular engineering. Single or a few electron transports or transfers include extremely low-energy logic switching or data storage. Molecular nanocomponents have great prospective in manufacturing low-power, ultradense, and low-cost computing chips [1].

These components primarily form self-assembled nanostructures which are having wide range of applications in nanochip technology (see **Figure 2**).

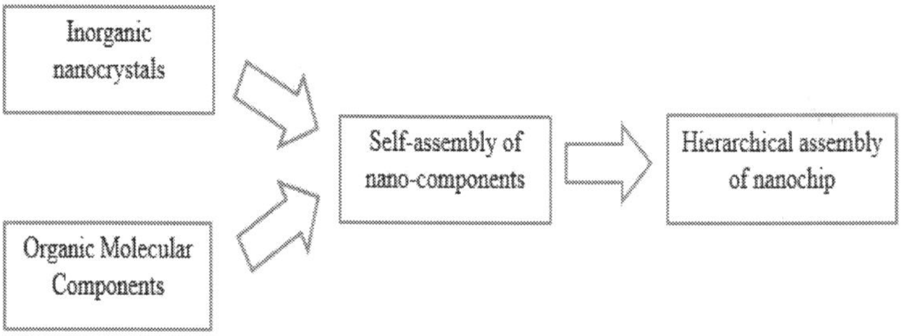

Figure 2. Nanoscale components for next-generation nanochips.

Among these kinds of materials, carbon-based materials such as CNTs, fullerenes, and graphene are gaining much consideration because of the ability of carbon to form bond with other elements which can easily be changed by physical interactions and chemical reactions.

4. Applications of nanoelectronics related to carbon-based materials

4.1. Thin-film transistors

In recent times, the performance of flexible thin film transistors (fTFTs) has been enhanced extraordinarily. CNTs have been used by numerous researchers for the fabrication of thin-film transistors (TFTs). As soon as the CNTs have been purified, solubilized, possibly functionalized, and finally deposited by one of the selective depositions or self-assembly methods, they can be associated in a device geometry which lets the study of their electronic properties. The attention in CNTs, especially SWCNTs, is because of their extraordinary electrical and structural properties [31]. The random linkage of SWCNTs is easily designed by drop, spin, and spray-coating methods. Although the SWCNTs having semiconducting properties show unipolar p-type behavior in an ambient environment in general [32,33], the random network of SWCNTs has revealed that this network can function as a semiconducting channel and conduction lane in the TFT device, as a result of dominant semiconductor and minimal density metallic tubes [34,35] (see **Figure 3**).

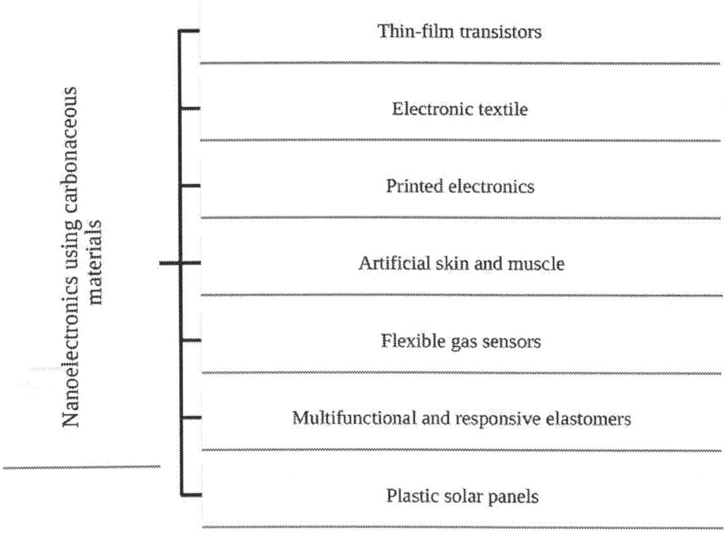

Figure 3. Applications of nanoelectronics in different fields.

Thin-film transistors based on controllable electrostatic self-assembled monolayer SWCNT network have been prepared by varying the density of nanotubes on the substrate made up of silicon. The densities of SWCNT network have been examined as a function of concentration and assembly time. It has been detected that the density of SWCNT network increases from 0.6 mm^2 to 2.1 mm^2, as the average on-state current (Ion) increases from 0.5 mA to 1.47 mA [36].

Organic thin-film transistors (OTFTs) with a spun cross-linked poly-4-vinylphenol (PVP) dielectric have been produced on a flexible substrate made up of polyethersulphone (PES). In order of increasing the electrical performance of OTFTs, a random network of SWCNTs was engaged as an underlay for carrier transfer devoid of sacrificing the flexibility of the TFTs [37].

SWCNT network (SWCNTN)-based thin-film transistors (TFTs) have been fabricated using roll-to-roll (R2R) gravure and inkjet on plastic foils to demonstrate the integrability in point of mass production with a low budget [38]. CNT-polyethylene terephthalate (PET) transistors have also been established by means of as-measured current gain cut-off frequency fT of 1 GHz and constant transconductance as high as 6 GHz [39].

It is worth observing that the greater mobility in combination with disappearing band gap means that graphene will be having less application for logic transistors which must be turned off, but it will be more suitable for high-frequency analog RF transistors where fTs in the order of 50 GHz are already probable. It has also been detected that at the nanoscale, the role of electrical contacts in electronic devices turns out to be increasingly important since they become an intrinsic part of the functional unit.

4.2. Artificial skin and muscle

The skin is the human body's largest organ which shields the body from disease and physical damage and helps to control body temperature. People with third-degree burns have to face a huge number of problems. They lose a huge proportion of their skin layers, for big wounds along with cases of venous ulceration. These are conditions where self-healing property of the skin lost normally [40–42]. Artificial skin and muscles are of great support in such kind of situations. Till now, the solitary other approach of covering up these parts is skin grafting, where the skin is taken from a different body part of the patient himself or from the donor's skin and grafted in the affected part. Such process is an expensive proposition and also involves a lot of uneasiness and pain. Scars are left on the treated wound and also the part from where the skin is taken. There is also the concern about the refusal of skin transplants from some other body parts. Bio-mimicking of artificial muscles or skins prepared from thin layers of polymers and CNTs is currently being fabricated by various scientific groups.

A rubberlike stretchable active matrix has been developed by means of elastic conductors [43] in which an ionic liquid of 1-butyl-3-methylimidazolium bis(trifluoromethylsulfonyl)imide was used and homogeneously distributed on single-walled carbon nanotubes (SWCNTs) as chemically unchanging dopants in a vinylidene fluoride-hexafluoropropylene copolymer matrix to produce a composite film. The active matrix sheet can be stretched biaxially and uniaxially by 70 % without mechanical or electrical damage. The elastic conductor helps in the production of electronic integrated circuits, which can be attached anywhere else, comprising random curved surfaces and movable portions, like the joints of a robot's arm.

High-resolution thin-film device has been synthesized to sense texture by means of touch. By using metal and semiconducting nanoparticles, a 100-nm-thick, large-area thin-film device was self-assembled such that the change in current density through the film and the light intensity (electroluminescent) was linearly proportionate to the local stress [44].

It is also extensively acknowledged that skin sensitivity is a significant concern for future robots used by humans in daily life for housekeeping and entertainment determinations. In spite of such fact, comparatively less progress has been made in the area dealing with pressure recognition compared to the areas of voice and sight recognition, primarily because good electronic artificial skin with mechanical flexibility and a large area is not yet obtainable. The construction of a delicate skin made up of thousands of pressure sensors involves a flexible switching matrix which cannot be realized with present silicon-based electronics. Within this perception, Someya and coworkers [43,45,46] have presented the incorporation of organic field-effect transistors and rubber pressure sensors, both of which being manufactured by low-cost processing technology such as large-area printing technology that can provide a perfect solution to realize a practical artificial skin.

4.3. Printed electronics (PE)

Printing is a multipurpose aiding technology for electronic products that cannot be prepared with the distinctive Si microelectronics technology. Polycrystalline and amorphous Si can make large-area products but have little carrier mobility and are stereotypically restricted to a rigid substrate. The printing of semiconductor and metals, conversely, permits the creation of large-area electronics on flexible substrates and allows high-volume scale economies. It is supposed that printed electronics will modernize our standard of living within the next two decades just as Si microelectronics has done in the past decades. The biggest segment of the total printed electronics manufacturing will be printed transistors and memory.

The PE printing technologies generally can be classified as either "mixed subtractive-additive" (where the processing steps include subtractive steps, such as etching or lift-off [47,48]) or "fully additive" (where the processing steps stringently include deposition only, devoid of etching or lift-off) methods. Printed electronics is generally based on polymers or organic molecules that are inexpensive and can be assembled into devices using room temperature fabrication methods. In recent times, a different conception, using nanoscale materials and solution-based deposition methods, is making inroads into the technological and profit-making sectors.

The impact of printed electronics on the recyclability of paper has been proved, and a case study has been done for smart envelopes in the field of courier and postal services [49]. The smart-printed envelope comprises of an amalgamation of both conventional (thin flexible batteries and resistors) and printed electronic constituents (conductive track layout based on nanosilver ink). For this purpose, a comparison among envelopes with and without these constituents (batteries, resistors, and conductive track layouts) was carried out through pilot-scale paper reprocessing tests. The generation of rejects in the course of the reprocessing process as well as the final quality of the recycled paper (mechanical and optical properties) were established and quantitatively assessed.

An effective and simplistic approach has been utilized to successfully synthesize approximately uniform silver nanoparticles (AgNPs) with particle size <10 nm, and the sintering of AgNPs has been demonstrated at room temperature for inkjet-printed flexible electronics [50].

Thin-film supercapacitors have been fabricated by means of materials which are printable to create devices having flexibility on plastic. The dynamic electrodes were prepared using sprayed SWCNTs' networks aiding as both charge collectors and electrodes. Using an aqueous gel electrolyte (printable) along with a liquid electrolyte (organic), the devices' performances illustrate very high energy and power densities (6 W h/kg for both the electrolytes and 23 and 70 kW/kg for aqueous gel electrolyte and organic electrolyte, respectively) which are analogous to presentation in other supercapacitor devices based on SWCNTs fabricated via different techniques [51].

In the field of flexible nanotube electronics, it has been revealed that polymer-supported network transistors show electronic properties that are similar to those of more orthodox, silicon-supported nanotube devices and are bendable and economical. Likewise, it is also demonstrated that nanotube networks could be made up as metallic networks by increasing the nanotube density. These could serve as economical, solution-treated contacts and interconnects [52].

4.4. Electronic textile

Electronic textile materials in the field of Nanoelectronics can be demarcated as fabrics where functions, interconnections, and electronics are intertwined into them. Electronic textiles give the conventional textile industry a novel additional value characterized by the possibility of making healthier everyday life, harmless and satisfied, conveying technological accomplishments closer to the people, through the apprehension of interfaces which are easy to use between humans and devices [53]. Clothings can now transform into various colors on command, give you a checkup, and connect by Wi-Fi. The first outcomes by now have appeared on store shelves, e.g., snowboarding jacket aimed to play MP3 music files with the assistance of controls stitched onto the jacket.

The main concern in this area is to assimilate flexible wires into textile substrate, connect them to electronics that can endure bending, twisting, and stretching, and power the whole ensemble. The requisite of upholding the textiles' mechanical flexibility lessens the possible technologies and materials appropriate for the objectives of the assignment: electrical conductivity along with its modulation cannot be attained through traditional semiconductors (like inorganic and silicon semiconductors). Certainly, a need for materials is there, whose mechanical properties do not fluctuate much from those of the textile substrates. Conductive polymers and CNTs, presently explored in the area of molecular electronics, are the unsurpassed contenders for undertaking such needs.

Flexible and wearable wire-shaped micro-supercapacitors have been designed which are based on highly aligned titania and carbon nanotubes [54]. A titanium wire has been used sheathed with outwardly aligned titania nanotubes as one of the electrodes to create all-solid-state micro-supercapacitors, in which the second electrode is carbon nanotube fiber or sheet. The capacitance of the subsequent micro-supercapacitor with a CNT sheet electrode was about three times of that for the analogous device with the second electrode based on a single CNT yarn. The exclusive wire-shaped structure made it promising for the wire-shaped micro-supercapacitors to be woven into various textiles and associated in series or parallel to meet a large variety of explicit energy demands.

A scalable nonliquid-crystal spinning procedure has been developed by researchers for the fabrication of continuous graphene fibers with tailored assembly for high-performance wearable supercapacitors [55]. These fibers infatuate surfaces with bark-like fine microstructure and differently designed cross sections with locally aligned dense pores, depending on the jet stretch ratio (R) throughout spinning. This nonliquid-crystal spinning approach could be stretched to the assembly of other two-dimensional nanomaterials into macroscopic fibers for uses in micro-devices, wearable electronics, and smart textile.

CNT-based interconnects could be fabricated by a number of methods. Two carbon nanotube (CNT) network fabrication procedures have been proposed, the normal spin rate coating (NR) and the slow spin rate coating (SR), and two interconnect assemblies, the single-layer structure

(SL) and the double-layer structure (DL), have been to make CNT interconnects [56]. A novel method has also been developed to produce nanocomposite textiles made from carbon nanotubes (CNTs) with improved sensing competencies [57]. Catering to the always increasing demand of upgraded sensors, this work deliberates the electrospinning fabrication scheme that has been engaged to cultivate novel CNT-based piezoelectric strain sensors. When these new piezopolymer composites which are based on CNTs were electro-spun into smart textiles, the ability of strain sensing (measured by voltage through the sensor) can be increased by 35 times, from 2.4 to 84.5 mV for the nanotubes with 0.05 wt%. The leading principle accountable for such enhancement was found to be the dipolar orientation in the piezoelectric material.

4.5. Flexible gas sensors

Sensing of gases signifies an essential subject in a huge variety of technological fields like processing in industries, storage, fuel cells, and separation. Conventional sensors for gases are made up, generally on substrates which are inorganic in nature, e.g., quartz, glass, and silicon wafers. The rigidity of such sensors related with the use of rigid substrates might restrict their use in a number of novel areas, e.g., aerospace science and portable devices, which demand lightweight and flexible sensing components. Flexible multi-sensors which support humidity, temperature, and detection of gas can be mass-produced at lesser cost and can consequently be incorporated onto Radio Frequency Identification tags or in smart textiles for logistic applications. CNTs which provide many sites where gasses can react, due to their extraordinary surface area and nanoscale assembly, set up a class of capable building blocks for manufacturing flexible chemical sensors on organic substrates.

Flexible vapor sensors using single-walled carbon nanotubes have been fabricated via the line-patterning technique [58]. Thin, strongly adhering films of single-walled carbon nanotube (SWCNT) bundles on flexible substrates such as polyethylene terephthalate (PET) have been used for sensing various kinds of vapors (hexane, toluene, acetone, chloroform, acetonitrile, methanol, water, etc.). These PET patterns are covered with films of electronically conductive SWCNT bundles (1–2 μm thick) by dip coating in aqueous surfactant-supported dispersions and attached in glass chambers fortified for sensing vapors. Flexible carbon nanotube sensors have also been developed for nerve agent simulators for the nerve agents sarin (diisopropyl methylphosphonate, DIMP) and soman (dimethyl methylphosphonate, DMMP) [59].

The fabrication of density-controlled single-walled carbon nanotube (SWCNT) thin films on poly(dimethylsiloxane) (PDMS) substrates has been reported by vacuum filtration and PDMS direct molding [60]. The flexible thin film of SWCNT indicates high stability (mechanically) with insignificant changes in conductance after twisting by 180°. This can be made known that conductance of SWCNT thin-film has high sensitivity to NH_3 gas partial pressure. It is also observed that the lesser the density of SWCNTs, the greater would be the sensitivity. The outcomes indicate that the flexible thin films of SWCNTs might be valid to future mobile sensors having high sensitivity.

High-performance hydrogen sensors have been manufactured with outstanding mechanical flexibility and durability on thin plastic sheets with the use of high-quality semiconducting single-walled carbon nanotubes adorned with discrete Pd nanoparticles [61]. The response times are normally less than 15 s for 1 % hydrogen, and the sensors are entirely recovered within 5 min in the air devoid of hydrogen. The flexible sensors can sense hydrogen with concentrations as low as 100 ppm (0.01 %). These flexible hydrogen sensors can find direct applications for sensing hydrogen leakage in various systems with demanding lightweight, mechanical flexibility, and high sensitivity.

A transparent and flexible glucose biosensor has also been reported on which multi-walled carbon nanotubes (MWCNTs) and glucose oxidase (GOx) are layer-by-layer (LBL) self-assembled onto a polymer-based substrate [62]. Biosensor exhibited a response range (linear) of 0.02–2.2 mM with a low detection limit of 10 μM. This significant act, pooled with the large-area preparation procedure, proves that this multilayer biosensor based on CNT is compatible for marketable applications.

The fabrication of flexible NO_2 sensors has also been done by layer-by-layer self-assembly of multi-walled carbon nanotubes and to study their gas-sensing properties [63]. First, a charged substrate (negative) is arranged by the construction of a monolayer which is organic (3-mercapto-1-propanesulfonic acid sodium salt; MPS) on a substrate of polyester with a pair of Au electrodes (comb-like). Afterward, two-cycle poly(4-styrene sulfonic acid-co-maleic acid)/poly(allylamine hydrochloride) (PSSMA/PAH) bilayers are placed on an MPS-modified substrate. Finally, multilayers of MWCNTs are designed as interchanging layers of cationic PAH and MWCNTs (negatively charged) on the modified substrate.

4.6. Multifunctional and responsive elastomers

The development in the field of nanoelectronics in the future relies on the multipurpose devices which can help in overcoming form-factor restrictions related with systems that only offer flexibility in combination with optical, electrical, and chemical properties. In that respect, polymer/CNT composites are among the maximum quoted contenders for nanoelectronics, a leading locus which stems mostly from their inherent electronic and structural properties.

The volume fraction-resolved, time-resolved, and frequency-resolved spectra of carbon black (CB) have been investigated at room temperature which is well above the glass transition temperature of the matrix-effective complex permittivity of well-characterized CB-filled ethylene butylacrylate copolymer samples which are surrendered to a uniaxial tension [64]. At low strain, the temporal evolution of permittivity while aging under stress showed a logarithmic growth phase followed by a logarithmic decay phase, whereas for sufficiently large strain, the permittivity behaviors overlap with those observed below the percolation threshold. An inspection of these materials using the surface and volume morphological evolutions under the act of a mechanical stress at the microscale by scanning electron microscopy (SEM) and atomic force microscopy (AFM) directed that aging during a few hundred hours, at a rational strain, generates cracks and voids that are aligned alongside the stretching direction.

The evolution of the absorption spectra for plastoferrites and carbon black-filled ethylene butylacrylate copolymer composites subjected to a uniaxial tension has been executed which is dependent on frequency and measured using a network analyzer (vector) as a function of the ratio of elongation over the frequency range (0.3–5 GHz) and at room temperature [65].

Loading a random network of SWCNT on a substrate (elastomeric) affords a two-terminal stretchable resistor having the capability to lodge strains more than 20 %. Such kind of strain leads to the distortion of individual SWCNTs within the whole network, in this manner varying their electronic properties in a reversible manner, because of the deviations either in the SWCNT-SWCNT contacts and/or band gaps [66,67]. This property can be used to make strain sensors with piezo-resistance gage factor, which can be defined as resistance modulation per strain, equivalent to those of the orthodox metal-strain gages.

4.7. Plastic solar panels

As the natural resources of the world are declining, the requirement of cheap and novel energy sources is developing. The sun is one such source which produces heat as well as light which can be harnessed and used for people's benefit. Solar cell designs are conventionally based on crystalline silicon, but not only that such material has been in short supply, the process of silicon solar is comparatively expensive too. As the solar industry struggles in order to reduce the cost to make it a feasible substitute of fossil fuels, many are now focusing toward plastic panels. In this viewpoint, photovoltaic devices which are flexible, e.g., polyimide solar cells, have recently been made up.

The photoelectrical properties of p-n junction organic solar cells have been investigated which are grounded on Cm and phthalocyanines mainly oxotitanium (IV) phthalocyanine (OTiPc) [68]. The Al/C and OTiPc/indium tin oxide (ITO) cells disclose relatively high-energy conversion efficiencies at 720 nm (10 p W cm^{-2}) due to the creation of a p-n junction between Cm and OTiPc. When ITO is used as the front electrode instead of Al, the stability of cells gets extraordinarily enhanced, so that the photocurrents remain unaffected for a long period of time even on white light illumination at high light intensity.

A pure carbon counter electrode (CE) has been manufactured for dye-sensitized solar cells (DSCs), by means of a substrate (industrial flexible graphite sheet) and activated carbon as the catalyst [69]. The CE presents series resistance (R_s) and charge-transfer resistance (R_{ct}) which are very low by merging the flexible graphite's high conductivity with the activated carbon's high catalytic property. The R_s and R_{ct} for the counter electrode are correspondingly only a quarter and two-thirds of those for a platinized tin oxide glass doped with fluorine (Pt/FTO). DSCs having cell areas of 0.15 and 1 cm^2 made up with this CE exhibit higher solar-to-electricity conversion efficiencies. The respective values are 6.46 % and 5 %, matching with 6.37 % and 2.91 % for the Pt/FTO-based devices.

The photovoltaic properties of indium tin oxide (ITO)/C_{60}/H_2 (pc), (pc = phthalocyanine)/Au-sandwiched solar cells have also been investigated [70]. The cell's photocurrent action spectra show that photocurrent which is generated at C_{60}/H_2 (pc) interface because of the diffusion of C60 excitons is due to the excited state of C60 that has a comparatively longer lifespan, while the cell made at a lower vacuum (331,025 torr) reveals a much lesser photocurrent (JSC51.4 μA cm^{-2}) since oxygen in C60 acts as carrier traps and increases the resistance of C60.

5. Conclusion

This chapter deals with the various approaches of nanotechnology toward nanoelectronics, viz., top-down and bottom-up technology in which bottom-up technology was found to be having superiority over the top-down approach for the fabrication of materials, circuits, etc. used in nanoelectronics. Further, this chapter shows the materials which can be used in nanoelectronics. Among the various materials used, carbon-based materials such as CNTs, fullerenes, and graphene are gaining much consideration because of the ability of carbon to form bond with other elements which can easily be changed by physical interactions and chemical reactions. Further, the recent advancements of C-filled plastics for uses in diverse types of optoelectronic, electronic, and sensor systems have been reviewed. Devices which are based on the plastic C/organic nanoelectronics are twistable, stretchable, and deformable into curvilinear shapes, in that way assisting the applications that would not be possible to achieve by using the brittle, rigid, and planar nature of electronics of today, e.g., Si or III–Vs. In order to build such robust and flexible electromechanical devices, multiple components with specific electrochemical and interfacial properties need to be incorporated into solo units. The chapter focuses on the application of carbon-based materials in various fields of nanoelectronics, viz., thin-film transistors, printed electronics, artificial skin and muscle, wearable electronics, flexible gas sensors, multifunctional and responsive elastomers, and plastic solar panels.

Author details

Gaurav Pandey, Deepak Rawtani* and Yadvendra Kumar Agrawal

*Address all correspondence to: rawtanid@gmail.com

Gujarat Forensic Sciences University, Gandhinagar, Gujarat, India

References

[1] B. Yu, M. Meyyappan, Solid-State Electron. 50 (2006) 536–544.

[2] S.P. Beaumont, Microelectron. Eng. 32 (1996) 283–295.

[3] International Technology Roadmap for Semiconductors, 2004 version (update).

[4] (a) A.K. Geim, K.S. Novoselov, Nat. Mater. 6 (2007) 183 (b) K.S. Novoselov, A.K. Geim, S.V. Morozov, D. Jiang, M.I. Katsnelson, I.V. Grigorieva, S.V. Dubonos, A.A. Firsov, Nature 438 (2005) 197.

[5] A.H. Castro Neto, F. Guinea, N.M.R. Peres, K.S. Novoselov, A.K. Geim, Rev. Mod. Phys. 81 (2009) 109.

[6] S. Ijima, Nat. (Lond.) 354 (1991) 56.

[7] M.A. Reed, C. Zhou, C.J. Muller, T.P. Burgin, J.M. Tour, Science 278 (1997) 252.

[8] X.D. Cui, A. Primak, X. Zarate, J. Tomfohr, O.F. Sankey, A.L. Moore, T.A. Moore, D. Gust, G. Harris, S.M. Lindsay, Science 294 (2001) 571.

[9] R. Compafi6, Nanotechnology 12 (2001) 85.

[10] J. Chen, M.A. Reed, A.M. Rawlett, J.M. Tour, Science 286 (1999) 1550.

[11] C.P. Collier, E.W. Wong, M. Belohradsw, E.M. Raymo, J.F. Stoddart, P.J. Kuekes, R.S. Williams, J.R. Heath, Science 285 (1999) 391.

[12] C.P. Collier, G. Mattersteig, E.W. Wong, Y. Luo, K. Beverly, J. Sampaio, E.M. Raymo, J.E. Stoddart, J.R. Heath, Science 289 (2000) 1172.

[13] S.J. Tans, A.R.M. Verschueren, C. Dekker, Nature 393 (1998) 49.

[14] S.J. Wind, J. Appenzeller, R. Martel, V. Derycke, P. Avouris, Appl. Phys. Lett. 80 (2002) 3817–3819.

[15] W. Liang, M.P. Shores, M. Bockrath, J.R. Long, H. Park, Nature 417 (2002) 725.

[16] J. Park, A.N. Pasupathy, J.I. Goldsmith, C. Chang, Y. Yaish, J.R. Petta, M. Rinkoski, J.P. Sethna, H.D. Abruna, P.L. McEuen, D.C. Ralph, Nature 417 (2002) 722.

[17] Y. Luo, C.P. Collier, J.O. Jeppesen, K.A. Nielsen, E. DeIonno, G. Ho, J. Perkins, H.R. Tseng, T. Yamamoto, J.E. Stoddart, J.R. Heath, Chemphyschem 3 (2002) 519.

[18] A. Bachtold, P. Hadley, T. Nakanishi, C. Dekker, Science 294 (2001) 1317.

[19] Y. Huang, X. Duan, Y. Cui, L.J. Lauhon, K.H. Kim, C.M. Lieber, Science 294 (2001) 1313.

[20] X.F. Duan, Y. Huang, C.M. Lieber, Nano Lett. 2 (2002) 487.

[21] Y. Chen, G.Y. Jung, D.A.A. Ohlberg, X. Li, D.R. Stewart, J.O. Jeppesen, K.A. Nielsen, J.F. Stoddart, R.S. Williams, Nanotechnology 14 (2003) 462.

[22] N.A. Melosh, A. Boukai, F. Diana, B. Gerardot, A. Badolato, P.M. Petroff, J.M. Heath, Science 300 (2003) 112.

[23] J.R. Heath, P.J. Kuekes, G.S. Snider, R.S. Williams, Science 280 (1998) 1716.

[24] F. Peper, J. Lee, S. Adachi, S. Mashiko, Nanotechnology 14 (2003) 469.

[25] H. Mizuta, S. Oda, Microelectron. J. 39 (2008) 171–176.

[26] M. Meyyappan, editor Carbon nanotubes: science and application. Boca Raton, FL: CRC Press; 2004.

[27] Sarazzin P. Chapter 8 in Ref. [21].

[28] J. Li, H.T. Ng, A.M. Cassell, W. Fan, H. Chen, Q. Ye, et al. Nano Lett. 3 (2003) 597–602.

[29] J. Li, Y. Lu, Q. Ye, M. Cinke, J. Han, M. Meyyappan, Nano Lett. 3 (2003) 929–933.

[30] Q. Ngo, B.A. Cruden, A.M. Cassell, G. Sims, M. Meyyappan, J. Li, et al. Nano Lett. 4 (2004) 2403–2407.

[31] A.P. Graham, G.S. Duesberg, R.V. Seidel, M. Liebau, E. Unger, W. Palmer, F. Kreupl, W. Hoenlein, Small 1 (2005) 382.

[32] A. Javey, J. Guo, Q. Wang, M. Lundstrom, H.J. Dai, Nature 424 (2003) 654.

[33] A. Javey, H. Kim, M. Brink, Q. Wang, A. Ural, J. Guo, P. McIntyre, P. McEuen, M. Lundstrom, H.J. Dai, Nat. Mater. 1 (2002) 241.

[34] X.Z. Bo, C.Y. Lee, M.S. Strano, M. Goldfinger, C. Nuckolls, G.B. Blanchet, Appl. Phys. Lett. 86 (2005) 182102.

[35] E.S. Snow, J.P. Novak, P.M. Campbell, D. Park, Appl. Phys. Lett. 82 (2003) 2145.

[36] X. Liu, C. Chen, P. Sharma, L. Zhang, Y. Wang, N. Hu, C. Song, L. Wei, Q. Liu, Y. Zhang, Physica E 74 (2015) 451–456.

[37] J.-H. Kwon, S.-I. Shin, M.-H. Chung, K.-Y. Dong, J.J. Pak, B.-K. Ju, Thin Solid Films 518 (2010) 6168–6173.

[38] J. Noh, M. Jung, K. Jung, G. Lee, S. Lim, D. Kim, S. Kim, J.M. Tour, G. Cho, Org. Electron. 12 (2011) 2185–2191.

[39] N. Chimot, V. Derycke, M.F. Goffman, J.P. Bourgoin, H. Happy, G. Dambrine, Appl. Phys. Lett. 91 (2007) 53111.

[40] R. Langer, J.P. Vacanti, Science 260 (1993) 920.

[41] R. Langer, J.P. Vacanti, Sci. Am. 273 (1995) 130.

[42] T. Mirfakhrai, J.D.W. Madden, R.H. Baughman, Mater. Today 10 (2007) 308. See also.

[43] T. Sekitani, Y. Noguchi, K. Hata, T. Fukushima, T. Aida, T. Someya, Science 321 (2008) 1468–1472.

[44] V. Maheshwari, R.F. Saraf, Science 312 (2006) 1501–1504.

[45] T. Someya, T. Sekitani, S. Iba, Y. Kato, H. Kawaguchi, T. Sakurai, Proc. Natl. Acad. Sci. U. S. A. 101 (2004) 9966.

[46] T. Someya, Y. Kato, T. Sekitani, S. Iba, Y. Noguchi, Y. Murase, H. Kawaguchi, T. Sakurai, Proc. Natl. Acad. Sci. U. S. A. 102 (2005) 12321.

[47] M. Guerin, A. Daami, S. Jacob, E. Bergeret, E. Benevent, P. Pannier, R. Coppard, IEEE Trans. Electron. Dev. 58 (2011) 3587–3593.

[48] X. Wei, G. Yang, U. Zschieschang, H. Klauk, B. Murmann, IEEE J. Solid-State Circuits 45 (2010) 1380–1388.

[49] http://dx.doi.org/10.1016/j.wasman.2015.01.005

[50] Deepak Rawtani, Y. K. Agrawal & Prajesh Prajapati, Interaction Behavior of DNA with Halloysite Nanotube–Silver Nanoparticle- Based Composite, BioNanoSci. (2013) 3:73–78.

[51] M. Kaempgen, C.K. Chan, J. Ma, Y. Cui, G. Gruner, Nano Lett. 9 (5) (2009) 1872–1876.

[52] K. Bradley, J.-C.P. Gabriel, G. Gruner, Nano Lett. 3 (2003) 10.

[53] http://www.gorix.com

[54] T. Chen, L. Dai, Energy Storage Mater. 2 (2016) 21–26.

[55] S. Chen, W. Ma, Y. Cheng, Z. Weng, B. Sun, L. Wang, W. Chen, F. Li, M. Zhu, H.-M.Cheng, Scalable non-liquid-crystal spinning of locally aligned graphene fibers for high-performance wearable supercapacitors, 15: 2015: 642–653.

[56] W.-C. Chiu, B.-Y. Tsui, Microelectron. Reliab. 54 (2014) 778–784.

[57] K. Laxminarayana, N. Jalili, Text. Res. J. 75 (2005) 670.

[58] K. Parikh, K. Cattanach, R. Rao, D.-S. Suh, A. Wu, S.K. Manohar, Sensors Actuat. B 113 (2006) 55–63.

[59] K. Cattanach, R.D. Kulkarni, M. Kozlov, S.K. Manohar, Nanotechnology 17 (2006) 4123–4128.

[60] C.-S. Woo, C.-H. Lim, C.-W. Cho, B. Park, H. Ju,D.-H. Min, C.-J. Lee, S.-B. Lee, Microelectron. Eng. 84 (2007) 1610–1613.

[61] Y. Sun, H. Hau Wang, M. Xia, J. Phys. Chem. C 112 (2008) 1250–1259.

[62] X.B. Yan, X.J. Chen, B.K. Tay, K.A. Khor, Electrochem. Commun. 9 (2007) 1269–1275.

[63] P.-G. Su, C.-T. Lee, C.-Y. Chou, K.-H. Cheng, Y.-S. Chuang, Sensors Actuat. B 139 (2009) 488–493.

[64] C. Brosseau, A. Mdarhri, A. Vidal, J. Appl. Phys. 104 (2008) 074105.

[65] C. Brosseau, W.N. Dong, A. Mdarhri, J. Appl. Phys. 104 (2008) 074907.

[66] T.W. Tombler, C.W. Zhou, L. Alexeyev, J. Kong, H.J. Dai, L. Lei, C.S. Jayanthi, M.J. Tang, S.Y. Wu, Nature 405 (2000) 769

[67] H. Maune, M. Bockrath, Appl. Phys. Lett. 89 (2006) 173131.

[68] H. Yonehara, C. Pat, Thin Solid Films 278 (1996) 108–113.

[69] J. Chen, K. Li, Y. Luo, X. Guo, D. Li, M. Deng, S. Huang, Q. Meng, Carbon 47 (2009) 2704–2708.

[70] K. Murata, S. Ito, K. Takahashi, B.M. Hoffman, Appl. Phys. Lett. 68 (3) (1996) 427.

Chapter 3

Copper-Indium-Gallium-diSelenide (CIGS) Nanocrystalline Bulk Semiconductor as the Absorber Layer and Its Current Technological Trend and Optimization

Nima Khoshsirat and Nurul Amziah Md Yunus

Additional information is available at the end of the chapter

http://dx.doi.org/10.5772/64166

Abstract

The Copper Indium Gallium diSelenide (CIGS) thin film solar cells are considered in this chapter. The interest in $Cu(In_{1-x}, Ga_x)Se_2$ thin film solar cells has increased significantly due to its promising characteristics for high performance and low cost. It is aimed to present an extensive evaluation on CIGS nanocrystalline bulk semiconductor and its application as an absorber layer for thin film solar cells. It is also aimed to improve the CIGS thin film solar cell efficiency through finding optimum ranges of material properties. The first section of this chapter gives an extensive overview on CIGS nanocrystalline bulk semiconductor background and technological trend. In the middle section, a brief review on CIGS Solar Cell processing and challenges are highlighted and the last section a numerical simulation results on the effects of each of constructive nano layer properties on cell performance are shown and compared with valid experimental results.

Keywords: copper-indium-gallium-diselenide (CIGS), indium sulfide (In_2S_3), nano-crystalline, thin film, solar cell, SCAPS

1. Introduction

Thin film solar cells are introduced and developed as the second generation of solar cells to provide high production capacity at lower energy and material consumption [1]. Main motivations for the growth of thin film photovoltaic (PV) are their potential for high-speed and high-throughput manufacturing and minimum material requirements that lead to cost reduction

[2]. This type of solar cell is made on cheap, large area and nonsilicon substrates such as glass, metal foil or plastic.

In 1976, three separate research groups, Panasonic Matsushita, University of Maine and RCA research group had formed the thin-film cell. Each group had used different absorber materials. Consequently, thin film solar cells were divided into three main subcategories based on their absorber materials such as amorphous silicon (a-Si), cadmium telluride (CdTe) and copper-indium-gallium-diselenide (CIGS). All these materials are direct-gap semiconductors that can absorb incoming solar radiation at a thickness much thinner than the required thickness for the silicon wafers in crystalline silicon (Si) solar cells. It is worth noting that the absorption of a-Si, CdTe and CIS/CIGS materials are significantly different. The highest value of solar radiation absorption belongs to CIS/CIGS that can absorb almost complete incoming radiation at first 3–4 μm of the material thickness and 95% of the radiation in its first 0.4–0.5 μm [3].

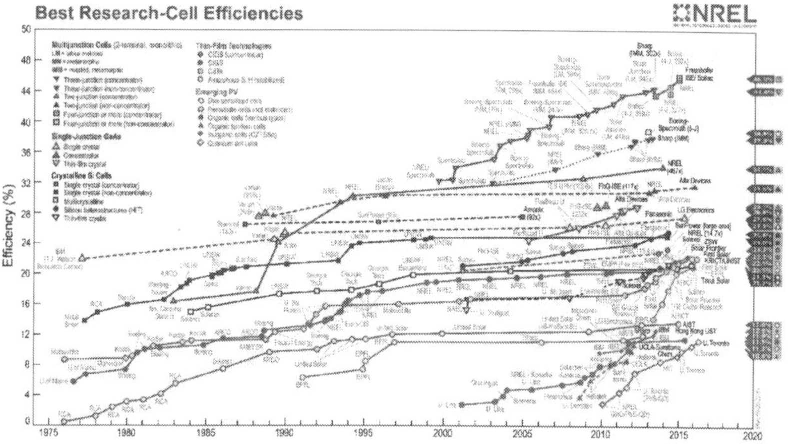

Figure 1. Best research solar cell efficiency released on April 2016 (Lawrence L. Kazmerski, National Renewable Energy Laboratory (NREL) [4].

Figure 1, which was recently published by US National Renewable Energy Laboratory (NREL), shows the best researched solar cell efficiencies that have been reported so far [4]. It can be observed that there is an efficiency gap between mono (single)-crystalline Si and poly-(multi)crystalline Si cells. Both of these silicon-based solar cells show higher levels of efficiency in comparison with other generations of solar cell. As a third-generation solar cell, the dye-sensitized cells have reached to over 10% efficiency for laboratory sample, but in general, their efficiency is less than other types of solar cells and are still known as an emerging PV technology. Among thin film solar cells, the CIGS cell has the highest record by the efficiency of over 20% (22.3% for a cell with glass substrate [4] and 20.4% for a cell with polyimide foil substrates [5]). The highest CdTe efficiency record is 22.1% [6], which is 0.2% lower than CIGS on glass substrate. Nevertheless, it needs to be studied and developed more in order to be successfully transferred from the laboratory stage to commercial stage and stay competitive in the PV market.

2. CIGS cell structure

Copper-indium-gallium-diselenide (CIGS) thin-film solar cells are multilayer thin film devices with $Cu(In_{1-x} Ga_x)Se_2$ nanocrystalline bulk semiconductor as the absorber material. The cheap substrate and monolithic interconnection of individual cells in a module are some initial advantages of CIGS thin film solar cells in comparison with silicon wafer-based solar cell. The suitable energy band gap of CIGS is another benefit of this compound semiconductor. Theoretically to ensure a sufficient absorption of the solar irradiation, the band gap of the absorber should be within the range of 1.0–1.8 eV with the optimum value of 1.5 eV [7]. The energy band gap of $Cu(In_{1-x} Ga_x)Se_2$ can be varied from 1.06 to 1.7 eV depending on the Ga/(Ga+In) ratio. Thus, the CIGS quaternary compound semiconductor is a good option to be used as an absorber material in solar cells.

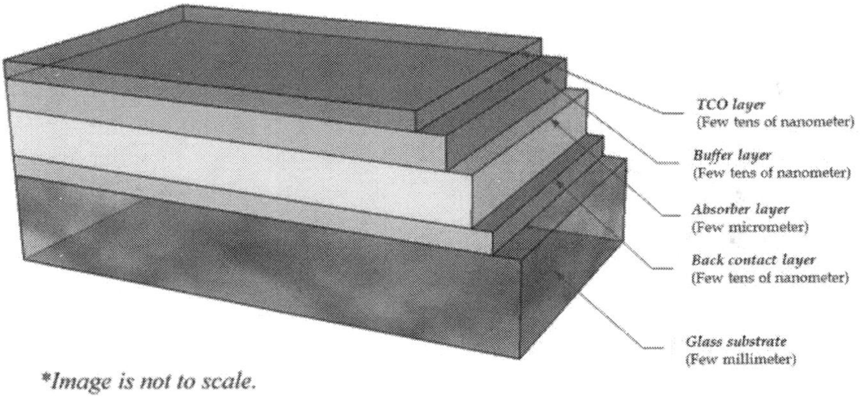

Figure 2. Typical CIGS thin-film solar cell structure [9].

Figure 2 shows the structure of CIGS solar cell. In this structure, the device is formed on low price substrates such as glass, polyimide foils, stainless steel, etc. The most common material used for substrate in CIGS technology is soda lime glass (SLG) due to its smooth surface, stability, electrical insulating features and more importantly its affordable price [8].

2.1. Back contact

The metal back contact is the first layer to be deposited on the substrate. This layer plays the role of an optical reflector as well as a contact for delivering the carriers to the load. Several materials have been proposed for the back contact such as molybdenum (Mo), tungsten (W), tantalum (Ta), manganese (Mn), titanium (Ti), etc. The most common material used for the back contact is molybdenum (Mo) (especially for soda lime glass substrates). This is because of molybdenum has low contact resistance to the absorber layer and it is stable at the processing temperature. These can limit the diffusion of atoms and hence minimize the destructive reactions during CIGS growth [10]. The process condition and growth parameters in the electrical and structural properties of Mo contact are important in the deposition method. For instance, although the resulting electrical resistivity of the Mo layer deposited at the low-pressure process is low, its adhesion to the substrate is poor and can be easily peeled off from the substrate's surface [11]. Thus, a compromise between the layer's adhesion and its resistivity is required during the back contact deposition.

2.2. Absorber layer

The absorber layer is deposited on the top of back contact layer as shown in **Figure 2**. $Cu(In_{1-x}Ga_x)Se_2$ is one of the most promising absorber material for PV applications. It is a direct band gap semiconductor. Its energy band gap (E_g) is within the range of 1.06–1.7 eV. The exact value of E_g depends on Ga/(Ga+In) ratio. The Cu concentration in the composition can also cause changes to CIGS band-gap value. A decrease in Cu concentration can cause an increase in band gap [12]. CIGS application in solar cells started with growth and the structural characterization of $CuInS_2$ (CIS) with the energy band gap of 1–1.06 eV in 1953. Then, some researches were conducted to increase the CIS band gap to the theoretical optimum value of around 1.4–1.5 eV. It was found that by adding gallium (Ga) to CIS will keep the overall number of group-III atoms, while In + Ga constant could lead to an increase in band gap [13]. This finding opens a new path for the enhancement of CIGS solar cell through adjusting the absorber layer band gap.

2.3. Buffer layer

The next layer in the CIGS solar cell structure is the buffer layer. This layer is deposited on the top of the absorber layer. The role of a buffer layer in the heterojunction is to form a junction with the absorber layer while leading maximum amount of incoming light to the absorber layer. The buffer layer should have a minimal absorption loss, low surface recombination and electrical resistance in photogenerated carriers driving out [14]. The most important features of the buffer layer are to protect the junction against chemical reactions and mechanical damage. It is also to optimize the band alignment of the cell, electrical properties and making a wide depletion region with p-type absorber layer. This will eventually minimize the carriers tunneling and maintain higher open circuit voltage value and later establish a higher contact potential [15]. In order to satisfy such desired features, the buffer layer should have a wider band gap in comparison with CIGS layer. In addition, the deposition process for the buffer layer should passivate the surface states of the absorber layer and provide a suitable conduction band alignment with the absorber to achieve higher efficiency [16]. The first experimental thin-film solar cell device using CIS absorber layer was an heterojunction between a p-type CuInSe layer and a thin layer of n-type CdS compound semiconductor. Further development of this structure was the vacuum evaporation of an undoped CdS layer, followed by indium-doped CdS layer to increase cell's performance and efficiency. Furthermore, the doped CdS would play the role of a transparent conductor layer [17]. The most efficient CIGS solar cell fabricated so far used CdS buffer layer. Although CdS buffer layer has yielded high-efficiency cell, its toxicity [18], incompatibility with in-line vacuum-based production method [19] and low-performance level of CdS/CIGS cells in short wavelength domain [20] led researchers to think about replacing CdS with a nontoxic material. At present, the development of Cd-free wide-band gap buffer layer is one of the main objectives in the field of CIGS thin film solar cells. The development of Cd-free device was initiated in 1992, and many different materials were proposed for it until now. These materials generally can be categorized in two main groups of zinc (Zn) based such as ZnS, ZnSe, ZnO, (Zn,Mg)O and indium (In) based including $In(OH)_3$, In_2S_3, In_2Se_3 [21].

2.4. Window layer

A thin layer of transparent conductive oxide (TCO) as front contact is just next to a buffer layer. This TCO layer should have a sufficient transparency to let most part of incoming light through the underlying layers. It must also have sufficient conductivity in order to transport the photogenerated current to the external circuit with minimum resistivity loss [22]. The most common used TCO material in CIGS solar cells is highly doped zinc oxide (ZnO) with the energy band gap of above 3.3 eV [23]. A doping of the ZnO layer is usually obtained by group III elements, exclusively with aluminum [24].

3. Criteria of material selection for the CIGS thin-film solar cell

There are some parameters that should be taken into consideration while choosing the material for different layers of a CIGS thin-film solar cell. The parameters such as the band gap, electron affinity, absorption coefficient, carrier concentration and many more can affect solar cell's characteristics and its output performance. The most effective parameters are the band gap and the electron affinity that will be discussed in the following sections.

3.1. Band gap

The most important parameter in CIGS solar cell is the absorber layer band gap. According to the theoretical considerations proposed by Loferski [25], the band gap values within the range of 1.4–1.5 eV is ideal for the absorber layer. **Figure 3** shows the maximum theoretical conversion efficiency as the function of semiconductor band gap for different absorber materials. Furthermore, the absorber layer should be a p-type and direct band gap semiconductor to have maximum conversion efficiency. This p-type semiconductor is proposed due to its longer electron diffusion length [26].

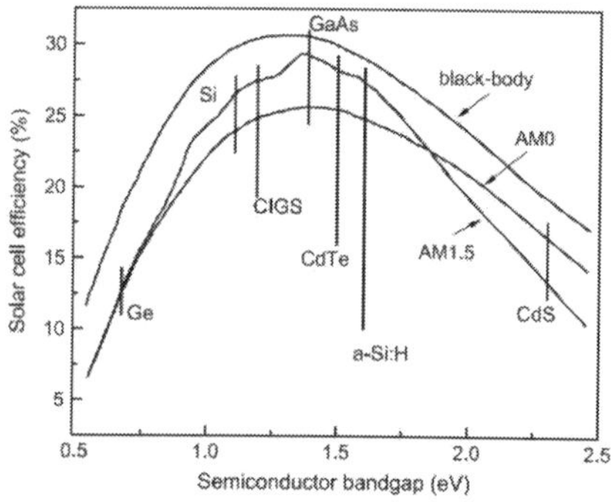

Figure 3. Theoretical maximum achievable efficiency versus absorber material band gap [25].

The energy band gap is also an important criterion for choosing a proper material for buffer and window layers. In this case, the energy band gap (E_g) needs to be adequately larger than absorber band gap to absorb as less photons as possible. It is assumed that photons with energies equal to and greater than the semiconductor material band gap can be absorbed in that material. In a CIGS solar cell, the light first passes through the window layer and those photons with energies more than the window layer's band gap will be absorbed in this layer. The remaining photons then pass through the buffer layer, and similarly, a part of incoming photons that have energies higher than the buffer band gap will be absorbed in the buffer layer. Most of the photogenerated carriers in the window and buffer layer cannot be collected due to their low mobility. **Figure 4** shows the photon flux at standard AM1.5 solar spectrum as a function of wavelength and maximum possible short circuit current density (Jsc) that can be generated in a solar cell as function of absorber material band gap. As it can be seen, the

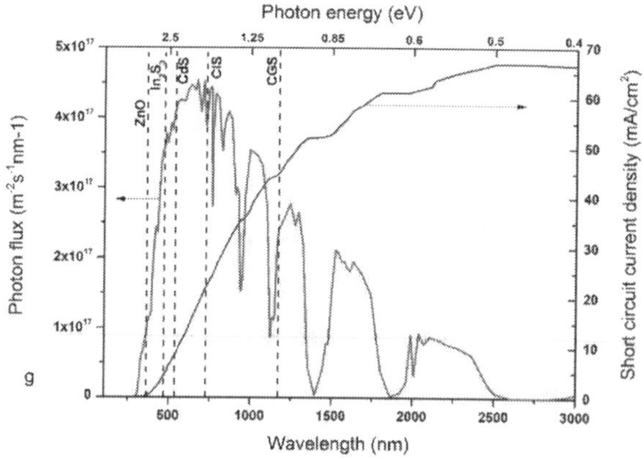

Figure 4. Photon flux at standard AM1.5 solar spectrum; maximum possible short circuit current density (Jsc).

photogenerated current loss is less than 1 mA/cm due to the absorption of photons in ZnO transparent conductive oxide (TCO) layer. Therefore, the loss of current caused by the absorption of light in the buffer layer is about 4 and 8 mA/cm^2 for In_2S_3 and CdS buffer materials, respectively.

In a cell with CIS absorber material in which its band gap is 1.06 eV as shown in **Figure 4**, the maximum photogenerated short-circuit current density (Jsc) is about 46 mA/cm^2, which is around 22 mA/cm^2 higher than maximum short-circuit current density in a cell with CGS absorber layer. Thus, although the band gap is not the only parameter, which can affect the cell performance, the selection of material for different layers in a solar cell should be under the band gap theoretical considerations in order to have the maximum absorption of light in absorber layer and minimum current loss.

3.2. Electron affinity

Another important parameter is electron affinity (X_e). The difference between electron affinity of absorber and buffer layer has an important role in the band alignment and shaping the discontinuity of energy band at the buffer/absorber interface. The discontinuity of conduction band at the interface that is called conduction band offset (CBO) can be positive (spike) or negative (cliff). This is due to the difference between absorber and buffer electron affinity. **Figure 5** shows the band alignment and formation of cliff and spike conduction band offset in CIGS solar cell. A cliff CBO at absorber/buffer interface can cause a reduction in open-circuit voltage (Voc) because of the lack of barrier height [27]. A positive CBO (spike) inhibits the flow of photogenerated carriers from the absorber to the buffer. A large spike makes a large barrier for carriers and therefore reduces the Jsc. However, a small spike does not act as a barrier [28]. Thus, the electron affinity of absorber and buffer layer should be compatible based on their energy band gap to maintain suitable levels of open-circuit voltage and short-circuit current density.

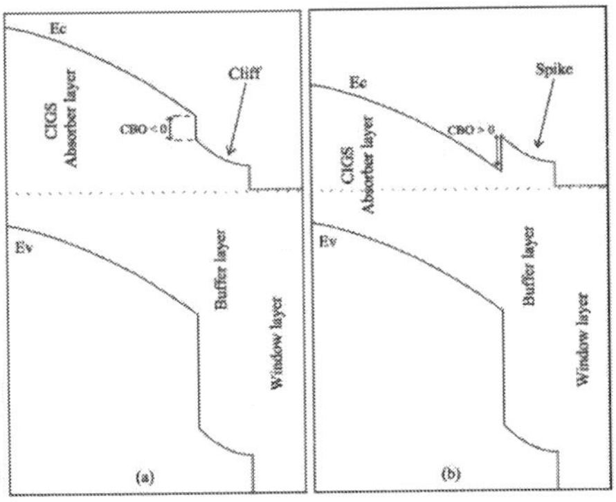

Figure 5. Band alignment between absorber and buffer layer. Conduction band offset: (a) Cliff and (b) Spike [20].

3.3. Quantum efficiency (QE)

Quantum efficiency is used as a tool for measuring the spectral response of the device. It gives a detail information about the absorption of photons and creation of carriers at different wavelength or photon energy levels. It is defined by the ratio of electrons collected from the device per incident photons at each wavelength:

$$QE\ (\lambda) = \frac{Number\ of\ collected\ electrons}{Number\ of\ incident\ photons} = \frac{I(\lambda)/q}{\varphi_p(\lambda)}$$

where the $I(\lambda)$ and $\varphi_p(\lambda)$ are photogenerated current and photon flow, respectively. Quantum efficiency is a relative value and its optimum number is 1 (i.e., 100%) for all wavelengths below the corresponding wavelength to the absorber band gap and zero for wavelength above it, but in reality, it is always less than 100%. **Figure 6** shows a typical CIGS cell's quantum efficiency (QE) curve and the loss mechanisms that can cause decreases in quantum efficiency. As shown, the reflection is one mechanism that decreases the quantum efficiency of the cell. It can be caused by light reflection at material interfaces or partial coverage cells' front surface that is made by front electrode.

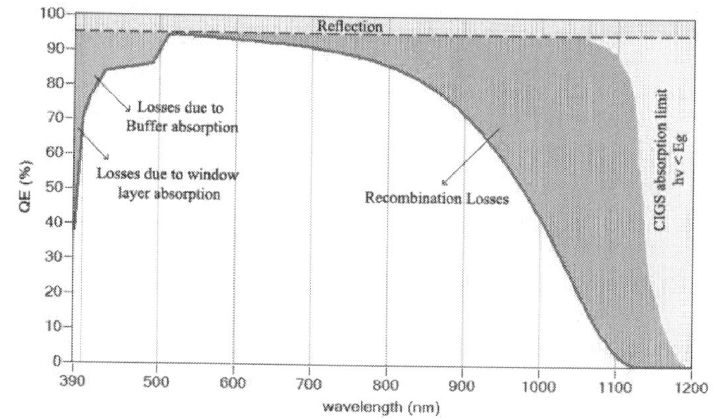

Figure 6. A typical cell's quantum efficiency curve and loss mechanism.

The other loss mechanism, which has destructive effect on cells' quantum efficiency, is the absorption of photons in the short-wavelength (UV) region. This arises from the absorption of light in buffer and window layer. The absorption in transparent conductive oxide window layer is typically low. This is due to the high-energy band gap of the materials that are usually used for this layer. But the buffer layer absorption is one of the major losses source in CIGS solar cells. As can be seen, there is a limitation at long wavelength because of the limit of the absorber layer absorption, which is based on its energy band gap. The short-circuit current and consequently the short-circuit current density of the cell can be obtained from the quantum efficiency:

$$J_{sc} = \int_{\infty}^{\lambda} G_{\lambda}(\lambda).QE(\lambda)d\lambda$$

where G_λ is the spectral irradiance of the reference distribution.

3.4. Recombination rate

The generation of carriers is counteracted by the carriers' recombination. The recombination phenomenon is the process, which acts to bring the solar cell back to equilibrium by the combination of exited electrons and holes. During the recombination, an electron relaxes back to the valence band from the conduction band by giving its energy to a photon or a phonon. While the electron transfers its energy to a photon in a single step, its energy will be divided and transferred to several phonons in several steps. There are three main mechanisms of recombination in a solar cell that can relax back the electrons from the conduction band to the valence band including radiative recombination, auger recombination and Shockley-Read-Hall (SRH) recombination [29]. The radiative mechanism happens when an electron gives its energy to a photon and emits it with this excess energy. Auger recombination mechanism includes the recombination of electron and hole and transferring the excess energy from the electron to the third carrier. The third recombination mechanism is Shockley-Read-Hall (SRH) recombination that includes emission of one or several phonons by transferring the excess energy of electron to them. In this mechanism, since the phonons' energy is ≤0.1 eV, band-to-band recombination requires simultaneous multiple phonon involvement. The SRH recombination usually involves defects states in the volume and at surfaces of the material. It is significant that the SRH recombination is the dominant recombination mechanism for CIGS solar cells due to the existence of defects in CIGS material. The recombination rate, R, is given by the classical Shockley-Read-Hall (SRH) description:

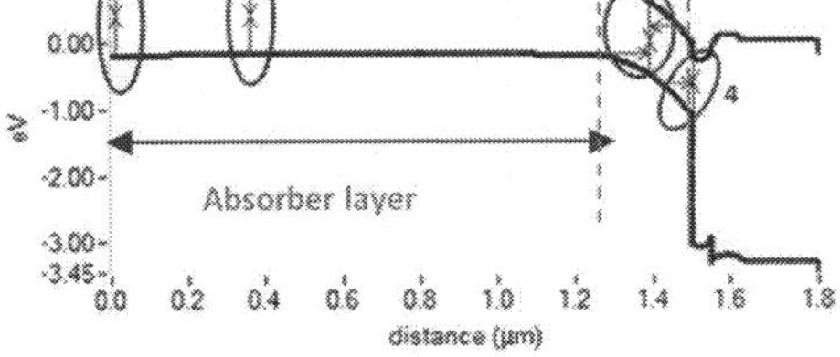

Figure 7. Band diagram and recombination paths in a CIGS solar cell: (1) Recombination at back contact. (2) Recombination at quasi neutral region. (3) Recombination in the space charge region (SCR). (4) Recombination at the interface.

$$R_{srh} = \frac{np - n_i^2}{\tau_{n_0}(n + n_1) + \tau_{p_0}(p + p_1)}$$

where the τ_{n_0} and τ_{p_0}, respectively, are the electron and hole minimum lifetime in the case of completely unoccupied defect states. The n_1 and p_1 are the electron and hole density, respectively, when the Fermi level lay at the energetic position of the defect. Furthermore, the recombination in thin film chalcopyrite solar cells (e.g., CIGS solar cells) can be categorized based on the place where it happens. There are four main recombination regions, including the recombination at the back contact (BC), quasi-neutral bulk, space charge region (SCR) and recombination at the interfaces (IFR). **Figure 7** represents the recombination regions in the band diagram of a CIGS solar cell. Region 1 represents the recombination at back contact and region 2 shows the quasi-neutral recombination (QNR) in the absorber layer. Region 3 is the place, where the recombination at space charge region (SCR) occurs, while region 4 is the place of recombination at the absorber/buffer interface happens. At each operation condition and cell structure, one of these recombination mechanisms is dominant. The basis of determination of the dominant recombination path in a CIGS solar cell is that, for all recombination paths, the diode saturation current density (*J*) can be written as below.

$$J_0 = J_{00} e^{(\frac{-E_a}{AKT})}$$

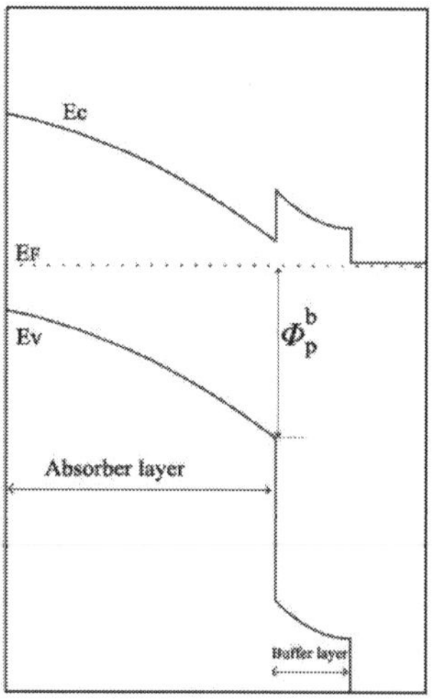

Figure 8. Interface barrier for holes at the absorber/buffer interface that is the energy distance between Fermi level and valence band edge at the absorber/buffer interface.

The J_{00} is the temperature-independent prefactor, E_a is the activation energy, "A" is the diode ideality factor that is the function of temperature and KT is the thermal energy. The most important parameter among these parameters is the activation energy that is dependent on the path of recombination. For the recombination in quasi-neutral balk (QNR) and recombination in space charge region (SCR), the activation energy (E_a) is equal to the energy band gap (E_g). In case of recombination at the absorber/buffer interface (IFR), E_a is equal to interface barrier for holes (φ_p^b). The interface barrier for holes normally is lower than the band gap. The (φ_p^b) that is shown in **Figure 8** is actually the energy distance between Fermi level (E_F) and valence band edge at the absorber/buffer interface.

4. A brief review on CIGS solar cell processing

CIGS thin-film solar cell fabrication process starts from the deposition of back contact layer onto the substrate. The condition of the process and growth parameters plays an important role in the electrical and structural properties of back contact layer. Therefore, the processing setup should maintain the desired resultant layers' features such as low resistivity and high adhesion to the substrate. The compatible deposition methods that are used to deposit molybdenum (Mo) as the common used back contact material are evaporation and sputtering [30, 31].

The second layer that should be deposited is the CIGS absorber layer. The essential criteria to select the processing technique of absorber layer are growth controllability, high deposition rate, reproducibility and low cost. Thus, several deposition techniques and growth methods are used for the absorber layer since the beginning of CIGS thin-film technology. There are advantages and disadvantages for each technique particularly for the criteria that are mentioned earlier. The CIGS deposition techniques can be divided into two main categories of high-cost vacuum-based technique and low-cost nonvacuum-based processing technique [32]. The vacuum-based deposition techniques, which are commonly used in CIGS deposition, are as follows [33]:

- Thermal evaporation such as coevaporation from pure elemental metals.
- Sputtering from metal selenide targets.
- Selenization of metal precursors.
- A hybrid of the above processes.

Although the vacuum deposition technique is applied to fabricate the commercialized CIGS cells and also the best laboratory-scale CIGS cell, there are some other low-cost CIGS thin film deposition techniques that are nonvacuum processes. Some of these techniques, which lead to fabricating a cell with an acceptable efficiency level, are the hydrazine-based deposition [34], screen printing [35], spray deposition [36], dip-coating [37], pulsed laser-assisted deposition (PLAD) [38], electrodeposition [39], etc. The considerable challenges in each alternative deposition and formation methods are producing CIGS layer, free from destructive concentrations of impurities and controlling the Ga concentration profile across the CIGS absorber. Both are effective factors on device efficiency [40].

The most commonly used deposition technique for the buffer layer in CIGS thin-film solar is the chemical-bath deposition (CBD) that is a low-cost, large-area process [41–43]. Since the chemical bath deposition technique is incompatible with in-line vacuum-based production method, which is used for commercialized CIGS solar cell fabrication in industry, severai alternative deposition techniques were proposed for buffer layer deposition process. These alternative methods are sputtering, atomic layer deposition (ALD), metal organic chemical vapor deposition (MOCVD), an ion layer gas reaction (ILGAR), a molecular beam epitaxy (MBE), etc. Each method may have some advantages and disadvantages, so a compromise between desired features and possible drawback is required. For instance, the atomic layer deposition (ALD) is a chemical vapor deposition method with very good thickness controllability and film uniformity; however, its weakness is the low growth rate in comparison with the other techniques [44]; while, the MOCVD is known as a fast and reliable deposition method. It can be integrated into in-line processes, but the MOCVD involve low controllability of impurity concentration across the layer. It is used for the deposition of zinc (Zn)-based buffer layers' deposition and the 13.4% efficiency was obtained from a $Cu(In,Ga)(S,Se)_2$ solar cell with a 20-nm MOCVD deposited ZnSe buffer layer [45].

Various deposition methods have been tested for TCO films such as RF or DC magnetron sputtering [46, 47], the sol-gel method [48], chemical vapor deposition (CVD) [49], pulsed laser deposition [50] and electrodeposition [51]. The TCO layers' deposition technique should be in low-temperature i.e. lower than 150°C. This is to avoid the detrimental interdiffusion across the underlying chalcopyrite layers. It also should be compatible with CIGS cells' in-line processing steps. Hence, magnetron sputtering of ZnO is the most commonly used deposition technique of TCO films among the above-mentioned processing methods. Magnetron sputtering is in a moderate price and consumes low temperature along with well controllable thickness and doping concentration. The RF magnetron sputtering is usually used for small area and laboratory-scale cell. But for the large-scale industrial production, the DC magnetron sputtering is used [52, 53].

5. CIGS thin-film solar cell challenge

Although the CIGS thin-film solar cell has already reached a technical maturity level that made it able to enter a mass production, there is still a large gap between the best commercial CIGS module efficiency of 12% [54] and the highest laboratory-scale efficiency of over 22%. Besides, there are some questions about the optimum cell structure, material properties and many more. These should be answered and explained in order to develop further the CIGS thin-film solar cell. Those unanswered questions arise obviously from an incomplete theoretical understanding about the cells.

In comparison with other types of solar cell, the CIGS cell is much more complex. This complexity comes from the nature of materials that are usually used in this type of solar cell.

Those materials are all compound semiconductors with tunable material properties. The other reason that makes the CIGS solar cells much more complex is the number of layers used in the cell structure. Thus, the complex fundamental semiconductor equations should be applied to all these layers with different material properties. That is why the study of this type of solar cell seems to be difficult and the theoretical understanding about some phenomenon in the CIGS cell's formation and operation are still under study. There are two major challenges that are briefly discussed below.

5.1. Defect nature

CIGS is a p-type semiconductor that is doped by intrinsic defects. In a CIGS bulk, there are vacancies such as V_{Cu}, V_{In}, V_{Se}, etc. and antisite defects like In_{Cu}, Cu_{In}, Ga_{Cu}, Cu_{Ga}. Some of these defects cause p-type doping such as V_{Cu} and some others may add n-type doping such as V_{Se} or $In_{Cu.}$ Theoretical and experimental studies have been done on CIGS intrinsic defects [55–57]. The results show that the copper vacancies dominate in CIGS and the p-type nature of this compound semiconductor arises from this defect. Although it is known that gallium (Ga) content can improve the CIGS electro-optical properties, but high Ga ratio could cause efficiency degradation. The physics behind this phenomenon is still under debate but one of the most possible reasons in antisite defects related to Ga. It is also identified that the diffusion of Na from the glass substrate to the absorber layer increases the carrier concentration in CIGS layer. This leads to a formation useful defect cluster such as the upgrading of CIGS material properties. However, more comprehensive understanding from the origin of defects and their contribution to CIGS electro-optical properties are still required.

5.2. Buffer layers

The highest demonstrated efficiency of CIGS solar cell has been obtained by a cell with a chemical bath-deposited (CBD) cadmium sulfide (CdS) buffer layer. Nevertheless, the cadmium is classified as a toxic material. Thus, a Cd-free buffer layer needs to be investigated and developed. Although, with the appropriate safety rules in fabrication process, the exposure to the cadmium can be avoided, due to breaking or disposal of cadmium containing products, the release of cadmium in the cell has a destructive effect on the environment and human health. Therefore, from July 1, 2006, two directives were regulated on the use of toxicants and heavy metals such as Hg, Cd, Pb in electronic products within the European Union: Directive on the Restriction of the Use of certain Hazardous Substances in Electrical and Electronic Equipment (RoHS) and Waste Electrical and Electronic Equipment Directive (WEEE) [58, 59]. Hence, in order to avoid the toxic heavy metal containing waste in solar cells, examining Cd-free buffer layers is necessary. Nontoxicity will ensure that stringent legislation relating to the use and disposal of toxic material cannot be an obstacle to improve the CIGS thin-film solar technology. In addition to the environmental problems, CdS buffer layer causes poor short-wavelength response due to absorption of the UV lights. This prevents high energy photons to reach the absorber layer and consequently reduces the quantum efficiency of the cell in UV region. Therefore, using an alternative buffer material or elimination of buffer layer in such a way can enhance the quantum efficiency with environmental friendly materials. These are other challenges of CIGS thin-film solar cells.

Absorber	Window	η (%)	Voc (v)	Jsc (mA/cm²)	FF (%)	Area (cm²)	Ref.
CuInS2	i-ZnO/ZnO/Al	13.5	0.604	30.6	73	25	[63]
CuInS2	i-ZnO/ZnO/Al	14.7	0.574	37.4	68.4	0.5	[64]
$Cu(In,Ga)Se_2^*$	i-ZnO/ZnO/Al	13.3	0.606	29.6	74	0.5	[65]
$Cu(In,Ga)Se_2$	i-ZnO/ZnO/Al	11.1	0.652	24.7	69.1	0.5	[66]
$Cu(In,Ga)Se_2$	i-ZnO/ZnO/Al	13.3	0.637	28.8	72.3	0.5	[67]
$Cu(In,Ga)Se_2$	i-ZnO/ZnO/Al	16.4	0.665	31.5	78	0.1	[68]
$Cu(In,Ga)Se_2$	i-ZnO/ZnO/Al	2.88	0.5136	30.83	47.65	3.75	[69]
$Cu(In,Ga)Se_2$	i-ZnO/ZnO/Al	12.1	0.66	26.9	68.4	25	[70]
$Cu(In,Ga)Se_2$	i-ZnO/ZnO/Al	12.3	0.525	31.8	73.6	0.5	[71]
$Cu(In,Ga)Se_2$	i-ZnO/ZnO/Al	14.1	0.648	34.3	63.3	0.5	[72]
$Cu(In,Ga)Se_2^*$	i-ZnO/ZnO/Al	12.4	0.556	31	72	6.25	[73]
$Cu(In,Ga)Se_2$	i-ZnO/ZnO/Al	10.8	0.592	29.5	62	900	[74]
$Cu(In,Ga)Se_2$	i-ZnO/ZnO/Al	12.9	0.662	26.8	72.6	900	[75]
$CuGaSe_2$	ZnO/Al	6	0.625	11.5	83	0.5	[76]
$CuGaSe_2$	ZnO/Al	3.9	0.785	14.5	34.3	0.5	[76]

Table 1. Summary of the Cd-free CIGS solar cell and modules with In_2S_3 buffer layer.

Development of Cd-free buffer layer for CIGS solar cell started in 1992 with an efficiency level of about 9% [60, 61]. Investigations during last decades show that some Cd-free material such as In_2S_3, ZnS and $Zn_{1-x}Mg_xO$ can potentially be used as alternative buffer layers in CIGS solar cells. The major advantage of these materials is that their band gap is larger than CdS band gap. It is worth noting that the In_2S_3 present a wide range of energy band gap from 2.1 to 2.9 eV [62]. In comparison with CdS, cells that use buffer materials with higher band gap have better spectral response in short wavelength due to less blue absorption loss in the buffer layer. **Table 1** shows the summary of Cd-free CIGS solar cells and modules that used In_2S_3 as one of the most promising alternative materials for the buffer layer.

The data shown in **Table 1** indicate that wide ranges of efficiency have been reported for the cells, which used same materials in absorber, buffer and window layer. These differences in efficiency value of the cells arise from differences in processing techniques, cells' geometrical and electro-optical properties of constituent layers. For instance, the cells that are shown with star mark (*) in **Table 1** are made by Cu(In, Ga)Se_2 absorber, In_2S_3 buffer and i-ZnO/ZnO/Al window layer. Both cells are made by same processing techniques but represent different efficiency (12.4 and 13.3%). The differences that are clearly reported for these samples are in

http://dx.doi.org/10.5772/64166

buffer layer thickness and band gap. The cell with the efficiency of 12.4% used a 50-nm buffer layer and the other one used 90-nm buffer layer. In one sample, they tried to control the oxygen content in buffer layer, and in another, they neglected it. These cells may have other differences in terms of absorber or window layer's properties. Thus, in parallel with ongoing efforts for mass production, it is necessary to optimize the cells in terms of layers' geometrical and materials' electro-optical properties in order to develop the CIGS thin film solar cells further. Solar cells' modeling and simulation are those beneficial approaches to reach an optimized cell. In other words, the cell performance can be simulated under different conditions by considering the independent and dependent parameters and the optimized cell can be concluded from superposition of simulation results.

6. Performance of the optimized CIGS solar cell through simulation

In this section, all the obtained optimized ranges that are reported in published results [9, 77–81] are used to evaluate the performance of a CIGS solar cell with optimum layer properties. This evaluation was done for the cells with uniform and graded band structure separately. **Figure 9(a)** shows the output parameters and the J–V characteristics of the cell with uniform band structure and optimized material properties, under dark and illuminated conditions. As shown, the efficiency of 20.16% is achieved through the simulation from this CIGS solar cell, which is higher than the best reported efficiency of a CIGS/In_2S_3/i-ZnO/ZnO/Al solar cell is 16.4% [68]. According to the simulation results and analysis, which are given in previous sections, the cell's open circuit voltage is enhanced by setting the absorber layer band gap at optimum value (1.2 eV). The cell's short-circuit current density is improved by setting the absorber layer thickness at optimum value (2±0.5 μm) and also by decrease in buffer and window layers' thickness that leads to the increase in absorber layer's absorption rate and decrease in buffer and window layers' recombination rate. The fill factor is also upgraded by setting the absorber layer band gap and thickness at optimum value and decrease in buffer

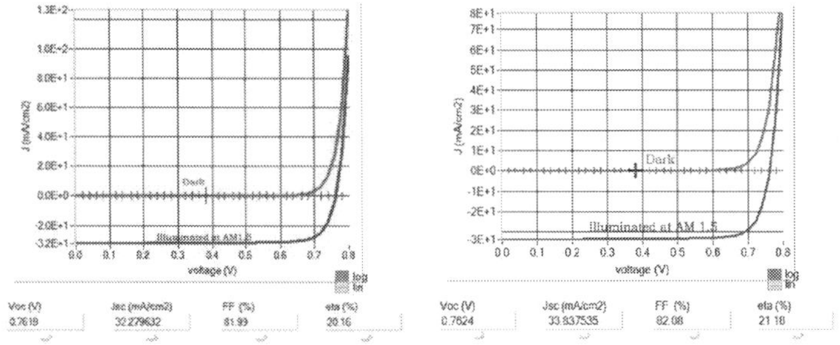

(a) (b)

Figure 9. J–V characteristics and the output parameters of the optimized cell under dark and illuminated conditions: (a) cell with uniform band structure, and (b) cell with graded band structure.

and window layers' thickness. A cell with optimum parameters and graded band absorber and buffer layer was also examined. The output characteristics of the cell with graded band structure and optimized material properties are shown in **Figure 9(b)**. As expected, it represents higher performance in comparison with the cell that has uniform band structure.

Cell	Material properties						Cells' output measurements			
	CIGS layer thickness (μm)	CIGS layer band gap (eV)	In_2S_3 layer thickness (μm)	In_2S_3 layer band gap (eV)	ZnO/Al layer thickness (μm)	ZnO/Al layer band gap (eV)	Voc (V)	Jsc (mA/cm²)	FF (%)	Efficiency, η (%)
Best cell reported in literatures [67]	2	N.A	0.03	2.7	0.1	N.A	0.665	31.5	78	16.4
Simulation results for the best cell reported in literatures	2	1.14–1.27	0.03	2.7	0.1	3.42–3.68	0.665	31.5	78	16.4
Simulated cell with uniform band structure	2	1.2	0.04	2.5	0.08	3.86	0.762	32.28	81.99	20.16
Simulated cell with graded band structure	2	1.2–1.7	0.04	2.5	0.08	3.86	0.762	33.84	82.08	21.18

Table 2. Summary of simulated cells' properties in comparison with the best reported cell.

Table 2 shows the summary of material properties that were used in simulated cells plus the best experimental and laboratory-scale CIGS/In_2S_3 cell's property, reported in the literature with their output measurements under the standard AM1.5 solar spectrum and environment temperature of 300°K. The differences in open-circuit voltage and the efficiency may arise from the difference in absorber and TCO layers' band gap, which is not reported for the experimental cell in the reference [68]. A cell with same parameter settings as used and reported in experimental reference was simulated and the results are also shown in **Table 2**.

As the proof of the present study, the best experimental cell with the settings that are mentioned in reference [68] was simulated to find approximate and possible ranges of unknown variables; CIGS layer and ZnO layer energy band gap. The simulation was performed by varying the CIGS and ZnO layers' band gap while the other independent parameters were kept constant. **Figure 10** shows the simulation results in the form of efficiency as the function of absorber and window layers' energy band gap in a color map.

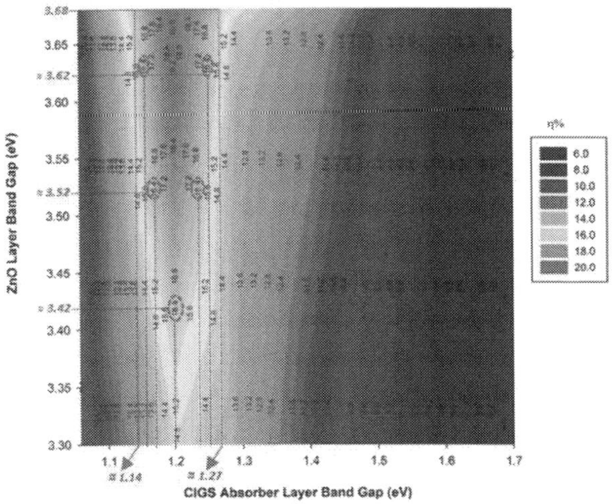

Figure 10. Simulation of best experimental cell: efficiency as the function of absorber and window layers' energy band gap.

Layer parameters	CIGS	In_2S_3	i-ZnO	n-ZnO
Thickness (μm)	2	0.04	0.07	0.08
E_g (ev)	1.2	2.5	3.3	3.68
X_e (ev)	4.25	4.25	4.6	4.24
ϵ_e	13.6	13.5	9	9
N_c (1/cm³)	2.2E+18	1.8E+19	2.2E+18	2.2E+18
N_v (1/cm³)	1.8E+19	4.0E+13	1.8E+19	1.8E+19
μ_n (cm²/Vs)	100	400	100	100
μ_p (cm²/Vs)	25	210	25	25
N_A (1/cm³)	1.0E+16	0	0	0
N_D (1/cm³)	0	1.0E+18	1.0E+16	1.0E+18

Table 3. Parameter set for the optimized CIGS/In_2S_3/ZnO solar cell.

According to the simulation results, the absorber and window layer energy band gaps of the best experimental CIGS/In_2S_3/i-ZnO/ZnO/Al solar cells with 16.4% output efficiency should have been in the range of 1.14–1.27 eV and 3.42–3.68 eV, respectively. In the color map, the cell performance was defined as an order of triple made of absorber layer's band gap, window layer's band gap and the cell's efficiency. Thus, each efficiency value in color map has specific corresponding value for absorber and window layer's band gap. Consequently, some particular points were specified in **Figure 10** with the efficiency value of 16.4%, equal to the efficiency reported for the cell in the reference. For example, if the measured value for the CIGS layer band gap in the best experimental cell is 1.2 eV, its window layer's energy band gap should be 3.42 eV and 16.4% efficiency. In other case, using absorber layer with 1.14 or 1.27 eV band gap for the cell would lead to having a window layer with energy band gap of 3.68 eV and hence obtains 16.4% efficiency from the cell. These outcomes can be considered as the proof of simulation result validity (**Table 3**).

7. Conclusion

CIGS thin film solar cells background, its technological trend and current challenges are presented in this chapter. The optimized CIGS/In_2S_3/i-ZnO/ZnO/Al solar cell material properties are proposed based on simulation results. As the proof of simulation results' validity, the best cells experimented and reported with their parameters are collected. These collections of cell information from literature are simulated and investigated. Some material properties from experimental cells are not reported directly in the literatures but have been calculated and proposed in a range through SCAPS simulation. The analyses, modeling and examination of the simulations results have shown that the optimization of material properties are promising and can improve the cell efficiency.

Author details

Nima Khoshsirat[1] and Nurul Amziah Md Yunus[2*]

*Address all correspondence to: amziah@upm.edu.my

1 Science and Engineering Faculty, Queensland University of Technology, Brisbane, QLD, Australia

2 Micro and Nano Electronic Systems Unit (MiNES) and Advanced Material Synthesis and Fabrication Laboratory (AMSF), Department of Electrical and Electronic Engineering, Faculty of Engineering, Universiti Putra Malaysia (UPM), Serdang, Malaysia

References

[1] J. Poortmans and V. Arkhipov, *Thin film solar cells: fabrication, characterization and applications*, vol. 5. John Wiley & Sons, West Sussex, England, 2006.

[2] J. W. Arnulf, "Progress in chalcopyrite compound semiconductor research for photovoltaic applications and transfer of results into actual solar cell production," *Solar Energy Materials and Solar Cells,* vol. 95, pp. 1509–1517, 2011.

[3] L. A. Kosyachenko, *Solar cells thin-film technologies*. Rijeka, Croatia: InTech, 2011.

[4] L. L. Kazmerski. *Best research solar cell efficiencies,* 2016. Available: http://www.nrel.gov/ncpv/images/efficiency_chart.jpg

[5] M. Powalla, W. Witte, P. Jackson, S. Paetel, E. Lotter, R. Wuerz, *et al.*, "CIGS cells and modules with high efficiency on glass and flexible substrates," *Photovoltaics, IEEE Journal of,* vol. 4, pp. 440–446, 2014.

[6] M. A. Green, K. Emery, Y. Hishikawa, W. Warta, and E. D. Dunlop, "Solar cell efficiency tables (version 48)", *Progress in Photovoltaics: Research and Applications,* vol. 24, pp. 905–913, 2016.

[7] N. H. Rafat and S. E. D. Habib, "The limiting efficiency of band gap graded solar cells," *Solar Energy Materials and Solar Cells,* vol. 55, pp. 341–361, 9/4/1998.

[8] S. Niki, M. Contreras, I. Repins, M. Powalla, K. Kushiya, S. Ishizuka, *et al.*, "CIGS absorbers and processes," *Progress in Photovoltaics: Research and Applications,* vol. 18, pp. 453–466, 2010.

[9] N. Khoshsirat, N. A. Md Yunus, M. N. Hamidon, S. Shafie, and N. Amin, "Analysis of absorber layer properties effect on CIGS solar cell performance using SCAPS," *Optik – International Journal for Light and Electron Optics,* vol. 126, pp. 681–686, 4/2015.

[10] L. Assmann, J. C. Bernède, A. Drici, C. Amory, E. Halgand, and M. Morsli, "Study of the Mo thin films and Mo/CIGS interface properties," *Applied Surface Science,* vol. 246, pp. 159–166, 2005.

[11] S.-Y. Kuo, L.-B. Chang, M.-J. Jeng, W.-T. Lin, Y.-T. Lu, and S.-C. Hu, "Effects of Growth Parameters on Surface-morphological, Structural and Electrical Properties of Mo Films by RF Magnetron Sputtering", *MRS Online Proceedings Library,* vol. 1123, pp. 1123–P05, 2008.

[12] W. Witte, R. Kniese, and M. Powalla, "Raman investigations of $Cu(In,Ga)Se_2$ thin films with various copper contents," *Thin Solid Films,* vol. 517, pp. 867–869, 11/28/2008.

[13] O. Lundberg, J. Lu, A. Rockett, M. Edoff, and L. Stolt, "Diffusion of indium and gallium in $Cu(In,Ga)Se_2$ thin film solar cells," *Journal of Physics and Chemistry of Solids,* vol. 64, pp. 1499–1504, 2003.

[14] B. McCandless and S. Hegedus, "Influence of CdS window layers on thin film CdS/CdTe solar cell performance," in *Photovoltaic Specialists Conference, Conference Record of the Twenty Second IEEE,* Las Vegas, NV, US, 1991, pp. 967–972.

[15] M. A. Contreras, M. J. Romero, B. To, F. Hasoon, R. Noufi, S. Ward, *et al.*, "Optimization of CBD CdS process in high-efficiency Cu(In,Ga)Se2-based solar cells," *Thin Solid Films*, vol. 403–404, pp. 204–211, 2002.

[16] U. Rau and H. W. Schock, "llc-4 – Cu(In,Ga)Se2 thin-film solar cells," in *Solar Cells*, M. Tom and C. Luis, Eds. Oxford: Elsevier Science, 2005, pp. 303–349.

[17] R.A. Mickelsen and WS Chen, "Development of a 9.4% efficient thin-film CuInSe2/CdS solar cell," presented at *the 15th IEEE Photovoltaic Specialists Conference*, New York, USA, 1981.

[18] S. Siebentritt, "Alternative buffers for chalcopyrite solar cells," *Solar Energy*, vol. 77, pp. 767–775, 2004.

[19] A. Romeo, M. Terheggen, D. Abou-Ras, D. L. Bätzner, F. J. Haug, M. Kälin, *et al.*, "Development of thin-film Cu(In,Ga)Se_2 and CdTe solar cells," *Progress in Photovoltaics: Research and Applications*, vol. 12, pp. 93–111, 2004.

[20] R. Klenk and M. C. Lux-Steiner, "Chalcopyrite based solar cells," in Thin Film Solar Cells: Fabrication, Characterization and Applications, *John Wiley & Sons, Ltd, West Sussex, UK*, pp. 237-275, 2006.

[21] D. Hariskos, S. Spiering, and M. Powalla, "Buffer layers in Cu(In,Ga)Se_2 solar cells and modules," *Thin Solid Films*, vol. 480–481, pp. 99–109, 2005.

[22] D. Ginley. and J. Perkins., "Transparent conducting oxides for advanced photovoltaic applications," *Photovoltaics International Journal*, 3rd edition, pp. 95–102, February 2009.

[23] M. Izaki and T. Omi, "Transparent zinc oxide films prepared by electrochemical reaction," *Applied Physics Letters*, vol. 68, pp. 2439–2440, 1996.

[24] T. Minami, H. Sato, H. Nanto, and S. Takata, "GROUP III Impurity Doped Zinc Oxide Thin Films Prepared By RF Magnetron Sputtering," *Japanese Journal of Applied Physics*, Part 2: Letters, vol. 24, pp. 781–784, 1985.

[25] J. J. Loferski, "Theoretical considerations governing the choice of the optimum semiconductor for photovoltaic solar energy conversion," *Journal of Applied Physics*, vol. 27, pp. 777–784, 1956.

[26] V. Avrutin, N. Izyumskaya, and H. Morkoç, "Semiconductor solar cells: recent progress in terrestrial applications," *Superlattices and Microstructures*, vol. 49, pp. 337–364, 2011.

[27] R. Scheer, "Activation energy of heterojunction diode currents in the limit of interface recombination," *Journal of Applied Physics*, vol. 105, p. 4505, 2009.

[28] A. Niemegeers, M. Burgelman, and A. De Vos, "On the CdS/CuInSe_2 conduction band discontinuity," *Applied Physics Letters*, vol. 67, pp. 843–845, 1995.

[29] L. A. Kosyachenko, "A Theoretical Description of Thin-Film Cu(In,Ga)Se2 Solar Cell Performance," in *Solar Cells - New Approaches and Reviews*, ed: Leonid A. Kosyachenko, INTECH, Croatia, 2015.

[30] Z.-H. Li, E.-S. Cho, and S. J. Kwon, "Molybdenum thin film deposited by in-line DC magnetron sputtering as a back contact for Cu(In,Ga)Se_2 solar cells," *Applied Surface Science*, vol. 257, pp. 9682–9688, 2011.

[31] M. Edoff, N. Viard, J. T. Wätjen, S. Schleussner, P.-O. Westin, and K. Leifer, "Sputtering of highly adhesive Mo back contact layers for Cu(In,Ga)Se2 solar cells," presented at the *24th European Photovoltaic Solar Energy Conference*, München, Germany, 2009.

[32] H.-W. Schock, "Properties of chalcopyrite-based materials and film deposition for thin-film solar cells," in *Thin-Film Solar Cells*, ed: Springer, Berlin, 2004, pp. 163–182.

[33] K. Seshan, Handbook of thin film deposition: *William Andrew, Elsevier*, UK, 2012.

[34] D. B. Mitzi, M. Yuan, W. Liu, A. J. Kellock, S. J. Chey, L. Gignac, *et al.*, "Hydrazine-based deposition route for device-quality CIGS films," *Thin Solid Films*, vol. 517, pp. 2158–2162, 2009.

[35] M. G. Faraj, K. Ibrahim, and A. Salhin, "Fabrication and characterization of thin-film Cu (In, Ga) Se2 solar cells on a PET plastic substrate using screen printing," *Materials Science in Semiconductor Processing*, vol. 15, pp. 165–173, 2012.

[36] D.-Y. Lee, S. Park, and J. Kim, "Structural analysis of CIGS film prepared by chemical spray deposition," *Current Applied Physics*, vol. 11, pp. S88–S92, 2011.

[37] M. Kaelin, D. Rudmann, F. Kurdesau, H. Zogg, T. Meyer, and A. N. Tiwari, "Low-cost CIGS solar cells by paste coating and selenization," *Thin Solid Films*, vol. 480, pp. 486–490, 2005.

[38] T. Nakada and S. Shirakata, "Impacts of pulsed-laser assisted deposition on CIGS thin films and solar cells," *Solar Energy Materials and Solar Cells*, vol. 95, pp. 1463–1470, 2011.

[39] V. S. Saji, I.-H. Choi, and C.-W. Lee, "Progress in electrodeposited absorber layer for $CuIn_{(1-x)}GaxSe_2$ (CIGS) solar cells," *Solar Energy*, vol. 85, pp. 2666–2678, 2011.

[40] Y. Hamakawa, Thin-film solar cells: next generation photovoltaics and its applications vol. 13: *Springer Science & Business Media*, Berlin 2004.

[41] Y. Hashimoto, "Chemical bath deposition of CdS buffer layer for CIGS solar cells," *Solar Energy Materials and Solar Cells*, vol. 50, pp. 71–77, 1998.

[42] A. Ennaoui, M. Weber, R. Scheer, and H. Lewerenz, "Chemical-bath ZnO buffer layer for CuInS 2 thin-film solar cells," *Solar Energy Materials and Solar Cells*, vol. 54, pp. 277–286, 1998.

[43] J. F. Trigo, B. Asenjo, J. Herrero, and M. T. Gutiérrez, "Optical characterization of In_2S_3 solar cell buffer layers grown by chemical bath and physical vapor deposition," *Solar Energy Materials and Solar Cells*, vol. 92, pp. 1145–1148, 2008.

[44] N. Naghavi, R. Henriquez, V. Laptev, and D. Lincot, "Growth studies and characterisation of In_2S_3 thin films deposited by atomic layer deposition (ALD)," *Applied Surface Science*, vol. 222, pp. 65–73, 2004.

[45] S. Siebentritt, P. Walk, U. Fiedeler, I. Lauermann, K. Rahne, M. C. Lux-Steiner, *et al.*, "MOCVD as a dry deposition method of ZnSe buffers for $Cu(In,Ga)(S,Se)_2$ solar cells," *Progress in Photovoltaics: Research and Applications*, vol. 12, pp. 333–338, 2004.

[46] L. Gao, Y. Zhang, J.-M. Zhang, and K.-W. Xu, "Boron doped ZnO thin films fabricated by RF-magnetron sputtering," *Applied Surface Science*, vol. 257, pp. 2498–2502, 2011.

[47] R. Menner, D. Hariskos, V. Linss, and M. Powalla, "Low-cost ZnO:Al transparent contact by reactive rotatable magnetron sputtering for $Cu(In,Ga)Se_2$ solar modules," *Thin Solid Films*, vol. 519, pp. 7541–7544, 2011.

[48] K.-m. Lin and P. Tsai, "Parametric study on preparation and characterization of ZnO:Al films by sol–gel method for solar cells," *Materials Science and Engineering: B*, vol. 139, pp. 81–87, 2007.

[49] D. Kim, I. Yun, and H. Kim, "Fabrication of rough Al doped ZnO films deposited by low pressure chemical vapor deposition for high efficiency thin film solar cells," *Current Applied Physics*, vol. 10, pp. S459–S462, 2010.

[50] L. Cao, L. Zhu, J. Jiang, R. Zhao, Z. Ye, and B. Zhao, "Highly transparent and conducting fluorine-doped ZnO thin films prepared by pulsed laser deposition," *Solar Energy Materials and Solar Cells*, vol. 95, pp. 894–898, 2011.

[51] J. Rousset, F. Donsanti, P. Genevée, G. Renou, and D. Lincot, "High efficiency cadmium free $Cu(In,Ga)Se_2$ thin film solar cells terminated by an electrodeposited front contact," *Solar Energy Materials and Solar Cells*, vol. 95, pp. 1544–1549, 2011.

[52] O. Kluth, G. Schöpe, B. Rech, R. Menner, M. Oertel, K. Orgassa, *et al.*, "Comparative material study on RF and DC magnetron sputtered ZnO: Al films," *Thin Solid Films*, vol. 502, pp. 311–316, 2006.

[53] M. Saad and A. Kassis, "Effect of rf power on the properties of rf magnetron sputtered ZnO: Al thin films," *Materials Chemistry and Physics*, vol. 136, pp. 205–209, 2012.

[54] R. Noufi, *CIGS PV technology – challenges, opportunities, and potential*. National Renewable Energy Laboratory. National Center for Photovoltaics, USA. 2013.

[55] Q. Cao, O. Gunawan, M. Copel, K. B. Reuter, S. J. Chey, V. R. Deline, *et al.*, "Defects in Cu (In, Ga) Se_2 chalcopyrite semiconductors: a comparative study of material properties, defect states, and photovoltaic performance," *Advanced Energy Materials*, vol. 1, pp. 845–853, 2011.

[56] J. Pohl, T. Unold, and K. Albe, "Antisite traps and metastable defects in Cu (In, Ga) Se_2 thin-film solar cells studied by screened-exchange hybrid density functional theory," *arXiv preprint arXiv:1205.2556*, 2012.

[57] B. Huang, S. Chen, H.-X. Deng, L.-W. Wang, M. A. Contreras, R. Noufi, *et al.*, "Origin of reduced efficiency in Cu (In, Ga) Se solar cells with high Ga concentration: alloy solubility versus intrinsic defects," *Photovoltaics, IEEE Journal of*, vol. 4, pp. 477–482, 2014.

[58] EC-2002a, "Directive 2002/95/EC of the European Parliament and of the Council of 27 January 2003 on the restriction of the use of certain hazardous substances in electrical and electronic equipment," *Official Journal of the European Union*, vol. L 37, pp. 19–23, 2003.

[59] EC-2002b, "Directive 2002/96/EC of the European Parliament and of the Council of 27 January 2003 on waste electrical and electronic equipment," *Official Journal of the European Union*, vol. L 37, pp. 24–38, 2003.

[60] R. Ortega Borges, D. Lincot, and J. Vedel, "Chemical bath deposition of zinc sulfide thin films," *Proceedings of the 11th EC PVSEC, Montreux, Switzerland*, pp. 862–865, 1992.

[61] C. Huang, S. S. Li, W. Shafarman, C.-H. Chang, E. Lambers, L. Rieth, *et al.*, "Study of Cd-free buffer layers using In x (OH, S) y on CIGS solar cells," *Solar Energy Materials and Solar Cells*, vol. 69, pp. 131–137, 2001.

[62] N. Barreau, S. Marsillac, D. Albertini, and J. Bernede, "Structural, optical and electrical properties of β-In2S3-3xO3x thin films obtained by PVD," *Thin Solid Films*, vol. 403, pp. 331–334, 2002.

[63] E. Yousfi, T. Asikainen, V. Pietu, P. Cowache, M. Powalla, and D. Lincot, "Cadmium-free buffer layers deposited by atomic later epitaxy for copper indium diselenide solar cells," *Thin Solid Films*, vol. 361, pp. 183–186, 2000.

[64] N. A. Allsop, A. Schönmann, H. J. Muffler, M. Bär, M. C. Lux-Steiner, and C. H. Fischer, "Spray-ILGAR indium sulfide buffers for $Cu(In,Ga)(S,Se)_2$ solar cells," *Progress in Photovoltaics: Research and Applications*, vol. 13, pp. 607–616, 2005.

[65] S. Gall, N. Barreau, F. Jacob, S. Harel, and J. Kessler, "Influence of sodium compounds at the Cu (In, Ga) Se 2/(PVD) In_2S_3 interface on solar cell properties," *Thin Solid Films*, vol. 515, pp. 6076–6079, 2007.

[66] D. Hariskos, R. Menner, S. Spiering, A. Eicke, M. Powalla, K. Ellmar, *et al.*, "In_2S_3 buffer layer deposited by magnetron sputtering for Cu (InGa) Se2 solar cells," in *Proc. Solar Energy Conference*, New Orleans, US, 2004.

[67] M. R. Hariskos D, Lotter E, Spiering S, Powalla M., "Magnetron sputtering of indium sulphide as the buffer layer in $Cu(InGa)Se_2$-based solar cells.," in *20th European Photovoltaic Solar Energy Conference, Barcelona, Spain*, 2005.

[68] N. Khoshsirat and N. A. M. Yunus, "Numerical Simulation of CIGS Thin Film Solar Using SCAPS-1D," presented at *the 2013 IEEE Conference on Sustainable Utilization and Development in Engineering and Technology*, Putrajaya, Malaysia, 2013.

[69] M. I. Hossain, "Fabrication and characterization of CIGS solar cells with In_2S_3 buffer layer deposited by PVD technique," *Chalcogenide Letters*, vol. 9, pp. 185–191, 2012.

[70] J. Sterner, J. Malmstr[??]m, and L. Stolt, "Study on ALD In_2S_3/Cu(In,Ga)Se_2 interface formation," *Progress in Photovoltaics: Research and Applications*, vol. 13, pp. 179–193, 2005.

[71] S. Spiering, L. Bürkert, D. Hariskos, M. Powalla, B. Dimmler, C. Giesen, *et al.*, "MOCVD indium sulphide for application as a buffer layer in CIGS solar cells," *Thin Solid Films*, vol. 517, pp. 2328–2331, 2009.

[72] H. A. Maksoud, M. Igalson, and S. Spiering, "Influence of post-deposition heat treatment on electrical transport properties of In_2S_3-buffered Cu (In, Ga) Se 2 cells," *Thin Solid Films*, vol. 535, pp. 158–161, 2013.

[73] S. Gall, N. Barreau, S. Harel, J. Bernede, and J. Kessler, "Material analysis of PVD-grown indium sulphide buffer layers for Cu(In, Ga)Se_2-based solar cells," *Thin Solid Films*, vol. 480, pp. 138–141, 2005.

[74] S. Spiering, D. Hariskos, M. Powalla, N. Naghavi, and D. Lincot, "CD-free Cu(In,Ga)Se_2 thin-film solar modules with In_2S_3 buffer layer by ALCVD," *Thin Solid Films*, vol. 431–432, pp. 359–363, 2003.

[75] S. Spiering, A. Eicke, D. Hariskos, M. Powalla, N. Naghavi, and D. Lincot, "Large-area Cd-free CIGS solar modules with In_2S_3 buffer layer deposited by ALCVD," *Thin Solid Films*, vol. 451, pp. 562–566, 2004.

[76] W. Vallejo, J. Clavijo, and G. Gordillo, "CGS based solar cells with In_2S_3 buffer layer deposited by CBD and coevaporation," *Brazilian Journal of Physics*, vol. 40, pp. 30–37, 2010.

[77] N. Khoshsirat, N. A. Md Yunus, M. N. Hamidon, S. Shafie, and N. Amin, "Analysis of absorber and buffer layer band gap grading on CIGS thin film solar cell performance using SCAPS," *Pertanika Journal of Science an Technology*, vol. 23, pp. 241–250, 2015.

[78] N. Khoshsirat and N. A. M. Yunus, "Numerical Simulation of CIGS Thin Film Solar Using SCAPS-1D," presented at *the 2013 IEEE Conference on Sustainable Utilization and Development in Engineering and Technology*, Putrajaya, Malaysia, 2013.

[79] N. Khoshsirat, N. A. Md Yunus, H. M. Nizar, S. Shafie, and N. Amin, "Optimization of CIGS Thin Film Solar Cells via Numerical Simulation," in *INTERNATIONAL SYMPOSIUM ON APPLIED ENGINEERING AND SCIENCES (SAES 2013)*, UPM, Serdang Malaysia, 2013.

[80] N. Khoshsirat, N. Md Yunus, M. N. Hamidon, S. Shafie, and N. Amin, "ZnO doping profile effect on CIGS solar cells efficiency and parasitic resistive losses based on cells

equivalent circuit," in *2013 IEEE International Conference on Circuits and Systems (ICCAS)*, Kuala Lumpur, Malaysia, 2013, pp. 86–91.

[81] J. Pettersson, T. Torndahl, C. Platzer-Bjorkman, A. Hultqvist, and M. Edoff, "The influence of absorber thickness on Cu (In, Ga) Se solar cells with different buffer layers," *IEEE Journal of Photovoltaics*, vol. 3, pp. 1376–1382, 2013.

Chapter 4

Epitaxial Cu_3Ge Thin Film: Fabrication, Structure, and Property

Fan Wu and Nan Yao

Additional information is available at the end of the chapter

http://dx.doi.org/10.5772/64060

Abstract

In this paper, the fabrication and electrical property characterization of epitaxial Cu_3Ge thin film are performed. By adjusting deposition parameters, the crystallinity of the as-grown Cu_3Ge thin films is improved, with the formation of twins within it. The average work function of epitaxial Cu_3Ge thin film is measured to be ~4.47 + 0.02 eV, rendering it a desirable mid-gap gate metal for applications in complementary metal-oxide semiconductor (CMOS) devices. The present study therefore shows an epitaxial Cu_3Ge thin film that is promising for applications.

Keywords: Cu_3Ge thin film, sapphire, twin, pulsed laser deposition, semiconductor metallization

1. Introduction

Cu_3Ge is a promising candidate and an alternative to Cu for contacts and interconnections in advanced integrated circuit devices. It has a relatively low bulk resistivity throughout the compositional window of 25–35 at% Ge [1, 2]. Its thermal stability against oxidation is also excellent [3, 4]. Furthermore, the service life of Cu_3Ge is considerably longer than Cu because the out-diffusion of Cu [4] is reduced. Not only different substrates have been used for growth of polycrystalline Cu_3Ge films (including GaAs [5–7], Si [4, 8–10], Ge [10, 11], $YBa_2Cu_3O_{7-x}$ [12], Si_xGe_{1-x} [8, 9, 13], Ta/TaN [14] and GaN [15]), but also various deposition methods have been used for Cu_3Ge thin film fabrication. For example, multiple physical vapor deposition methods, such as electron beam deposition, sputtering [1, 2, 4, 11, 12, 16] and thermal evaporation [17], or chemical approaches such as vapor-solid reaction [14], have been exploited. Despite the achievements in this field, the Cu_3Ge films reported up to now were mostly polycrystalline with impurity phase [1,2,4–13]. Consequently, Cu_3Ge films with better crystallinity are highly desired, in order to minimize diffusion paths (grain boundaries) and lower electrical resistivity.

In this chapter, we show the fabrication of epitaxial Cu_3Ge thin films with significantly improved crystallinity, due to a modified deposition route of Cu_3Ge thin films. The average work function of epitaxial Cu_3Ge thin film is measured to be ~4.47 + 0.02 eV, rendering it a desirable mid-gap gate metal to be used for applications in CMOS devices. Since sapphire has been widely used as a substrate in high-power and high-frequency CMOS integrated circuits, especially for high-power radio-frequency applications, the integration of Cu_3Ge as contacts or interconnections onto sapphire substrate will have potential applications in the semiconductor industry.

2. Experimental procedure

Pulsed laser deposition (PLD) was used to deposit Ge and Cu thin films repetitively on sapphire substrates. For Cu atoms to mix and bond with Ge atoms intimately, ultrathin Ge and Cu layers were deposited in each repetition. Cu_3Ge film is thus expected to have improved crystallinity as laser ablation provides excess kinetic energies of Cu/Ge atoms. A series of five Cu_3Ge films are fabricated at 400 ± 10°C, with systematically changed deposition parameters for investigating their correlation with the crystallinity of Cu_3Ge films. Specifically, 90 repetitions of Ge and Cu layers are deposited for all five samples, with a changing pulse number of Ge and Cu deposition for sample 1–5. For example, 35 pulses of Cu and five pulses of Ge are deposited in one repetition for sample 1, and a total of 90 repetitions are performed. Similarly, pulses of Cu in one repetition for sample 2–5 are 25, 15, 14, and 7, while pulses of Ge in one repetition for sample 2–5 are 5, 5, 2, and 1. Before deposition, multi-step cleaning is performed on the substrates. They are first cleaned with boiled acetone for 5 min, and then ultrasonically cleaned in acetone and methanol for 5 min each. Nitrogen gun is used to dry the cleaned substrates, which are then mounted into the chamber 4 cm away from the target. The pulsed KrF excimer laser used has a wavelength of 248 nm and pulse duration of 25 ns. The pre-deposition vacuum is $\sim 10^{-7}$ Torr, while the actual deposition vacuum is $\sim 3.0 \times 10^{-4}$ Torr due to heating. The laser source provides a laser with constant exciting voltage (23.8 keV [18–26]), and hits the pure Cu and Ge targets (from ESPI) at an incidence angle of 45°. Since the spot size of the landing laser is ~2 mm × 3 mm and the landing laser beam energy is ~0.29–0.30 J, the laser beam energy density is ~4.8–5 J cm^{-2}.

High-resolution transmission electron microscope (HRTEM) JEOL-2010F with a point-to-point resolution of 0.18 nm is used to characterize the microstructure of the deposited films. Focus ion beam (FIB) is used for preparing TEM samples. Rigaku X-ray diffractometer (XRD) is used to perform XRD θ–2θ scan analysis, with Cu-Ka X-ray source. Atomic force microscopy (AFM) is exploited to obtain height images for characterization of thin film morphology. All AFM scans are performed in tapping mode, using cantilevers with a resonance frequency close to 330 kHz. The AFM image resolution is 512 × 512 pixels, and scan rate is 1.5 × 1.5 μm^2. Gwyddion is used to process the obtained AFM images.

The local work function of epitaxial Cu_3Ge thin film was measured by Kelvin probe force microscopy (KPFM). In KPFM measurements, the conductive tip (NSG03/Pt) was working in tapping mode and lift mode scan. Reliable topographic images were obtained before the KPFM measurements. In KPFM tests, an AC and a DC bias were applied on the cantilever while the sample remained grounded. The frequency of the AC signal was set at ~2kHz lower than the resonance frequency of the cantilever and the amplitude AC voltage was 560 mV. The lift scan height was 50 nm and the scan rate was 0.75 Hz.

3. Results and discussion

To characterize the microstructures of the deposited thin films, XRD θ–2θ patterns are obtained and shown in **Figure 1**. Since Cu/Ge layer thickness in each repetition is decreasing from sample 1 to sample 5, the crystallinities of Cu_3Ge films are improved. The two peaks corresponding to c-sapphire (0 0 6) and (0 0 1 2) planes exist in all XRD θ–2θ patterns, showing the substrate orientation. The film of sample 1 is polycrystalline Cu_3Ge with impurity Ge phase, as evidenced by the peak at 27.32° belonging to Ge (1 1 1) plane, and multiple peaks belonging to Cu_3Ge. Note that for sample 1, the deposited Cu/Ge layers in one deposition repetition were too thick to react with each other completely, such that extra Ge phase exists. The impurity Ge phase also exists for samples 2 and 3 (as evidenced by the peak at 27.32° for Ge (1 1 1)), due to the high Ge:Cu pulse ratio in each repetition. Furthermore, the crystallinities of Cu_3Ge films are also improved for samples 2 and 3, as evidenced by the disappearance of peaks belonging to Cu_3Ge (-1 3 1) and (0 0 4) planes. For samples 4 and 5, no impurity Ge phase exists, due to the reduced pulse numbers of Ge/Cu. More importantly, the crystallinities of Cu_3Ge films are

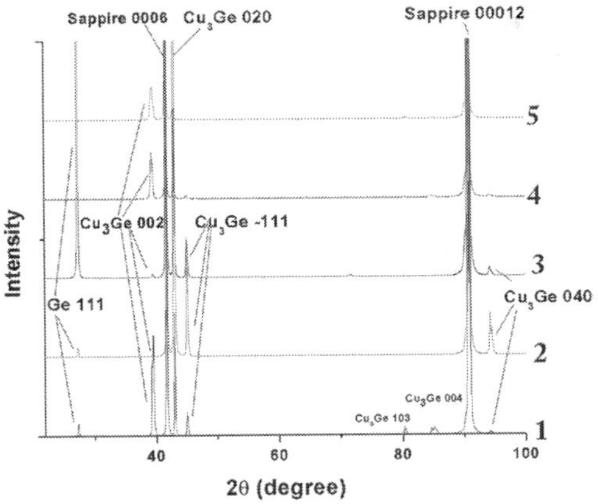

Figure 1. XRD θ–2θ patterns for samples on *c*-plane sapphire (Al_2O_3{0 0 1}). For all five samples, the laser shot frequency is 1 Hz, the deposition temperature is 400°C, and 90 repetitions of depositions of Cu/Ge were performed, while numbers of laser pulses for Cu and Ge depositions are different: sample 1: Cu:Ge=35:5; sample 2: Cu:Ge=25:5; sample 3: Cu:Ge=15:5; sample 4: Cu:Ge=14:2; and sample 5: Cu:Ge=7:1.

significantly improved with only two Cu_3Ge peaks remaining, proving that bi-epitaxial Cu_3Ge thin films have been grown on c-substrate.

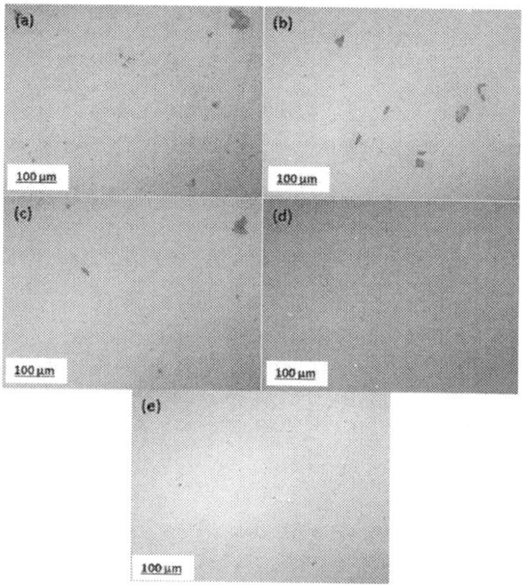

Figure 2. (a)–(e) The plan-view images of samples (1)–(5) respectively, observed by optical microscope.

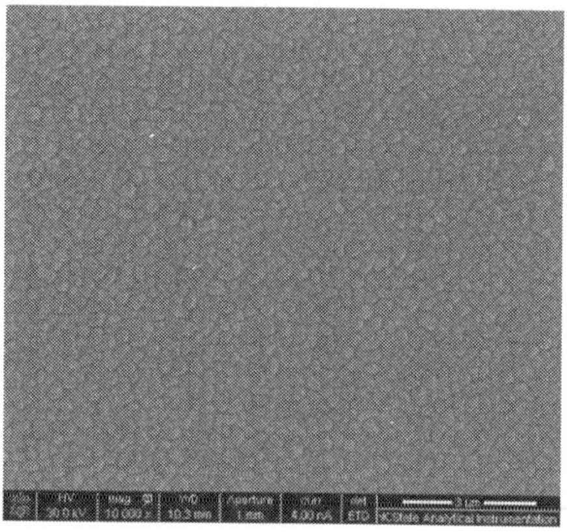

Figure 3. Typical plan-view SEM image of the Cu_3Ge film in sample 5, showing a uniform Cu_3Ge film grown via 3-D growth mode, without the formation of chunks and excessive Ge grains.

The quality and crystallinity of Cu_3Ge films in sample 1–5 are also demonstrated by the plan-view images of optical microscope images, shown in **Figure 2**. **Figure 2(a)–(e)** correspond to the plan-view images of samples 1–5, respectively. Chunks appear in the plan-view images of samples 1, 2, and 3, possibly due to the existence of excessive Ge grains in addition to the Cu_3Ge phase. In contrast, uniform films are observed in **Figure 2(e)** and **(f)**, proving the high quality of the Cu_3Ge films with good crystallinity for samples 4 and 5. SEM and AFM analyses are performed to show more detailed surface morphology of sample 5. The plan-view SEM image (**Figure 3**) shows the surface morphology of Cu_3Ge film, with an average diameter of Cu_3Ge islands to be ~300–500 nm. The complementary AFM height map (**Figure 4**) demonstrates the average heights of Cu_3Ge islands to be ~180–240 nm.

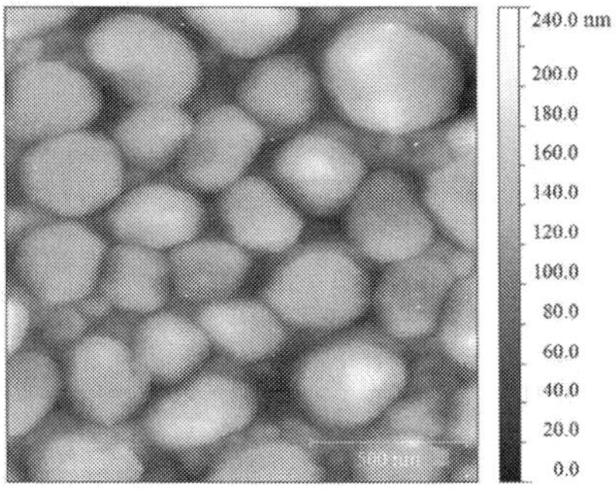

Figure 4. Typical AFM image of Cu_3Ge film on c-sapphire in sample 5, showing the diameters of these islands to be ~300–500 nm, with heights ranging from 180 to 240 nm.

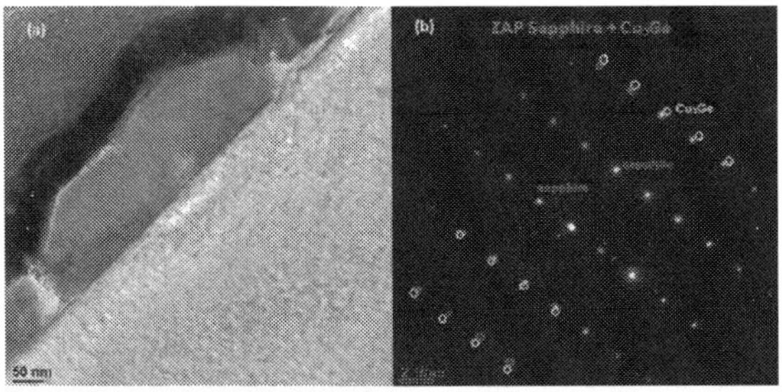

Figure 5 (a) A typical bright-field cross-section TEM image showing the film morphology of sample 5 and (b) the [1 1 0] zone-axis pattern (ZAP) of sample 5, clearly demonstrating an epitaxial relationship between the single-crystal Cu_3Ge island and c-sapphire substrate [42].

TEM is performed to further study the epitaxial film morphology, as demonstrated in **Figure 5(a)**. Note that the Au/Pt layers on top of Cu_3Ge are deposited during ample preparation for protection. Based on the morphology of the Cu_3Ge film, it may grow on the *c*-plane sapphire via 'layer-plus-island growth' mode following a two-step process. Initially, the Cu_3Ge grew in a layer-by-layer fashion on the *c*-plane sapphire to form an ultrathin uniform film up to several monolayers thick; beyond a critical thickness, the Cu_3Ge growth continued through the nucleation and coalescence of adsorbate islands. A sharp interface without secondary phases is observed between c-sapphire and epitaxial Cu_3Ge film, over a large area. Selected area diffraction pattern (SADP) analysis (**Figure 5(b)**) was performed at the Cu_3Ge/sapphire interface for studying orientation relationships. Overlapping of the low-order diffraction points of film and substrate corroborates their epitaxial relationship. A careful examination does show the splitting of higher order diffraction points of two phases.

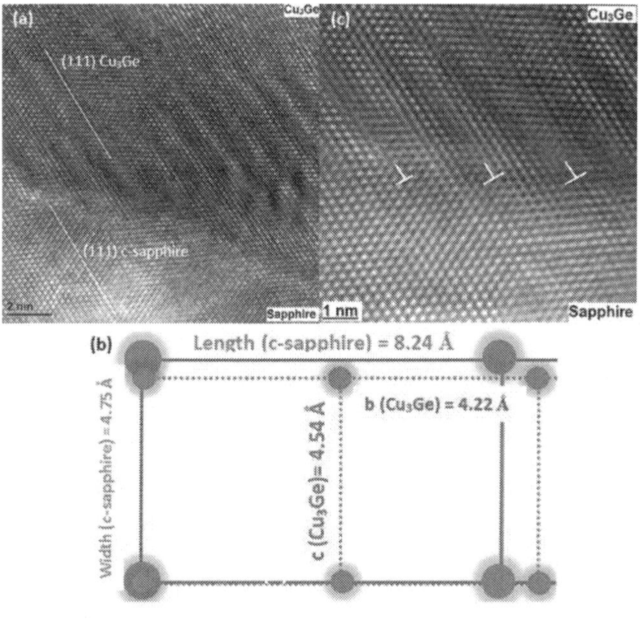

Figure 6 (a) An HRTEM image showing the atomic structure at the interface region between the Cu_3Ge island and c-sapphire substrate. The viewing direction is [1 1 0] zone of c-sapphire ([11–20] zone in Miller-Bravais Indices system), (b) the schematic illustration of the matching scenario at the interface between e_1-Cu_3Ge (0 1 0) and c-sapphire (0 0 1), and (c) the HRTEM image at a higher magnification showing the atomic structure at the interface in detail, at the region close to that of (a) [42].

HRTEM investigation was performed to study the atomic structure at the Cu_3Ge/c-sapphire interface, as shown in **Figure 6(a)**. The Cu_3Ge/c-sapphire interface is semi-coherent [27], due to visible periodic contrast along the interface induced by misfit dislocations, which accommodates the misfit strain between the two phases. For c-sapphire and Cu_3Ge, the interface planes are (0 0 1) and (0 1 0) planes respectively, and the viewing direction is [1 1 0] zone for both phases. **Figure 6(b)** shows the schematic illustration of the matching scenario at the interface. The atoms of sapphire within each lattice on (0 0 1) plane are located at the four corners of a rectangle (width = 4.759 Å, length = 8.243 Å), and the lattice of (0 1 0) plane Cu_3Ge

http://dx.doi.org/10.5772/64060

is also in rectangular shape (width = 4.54 Å, length = 4.22 Å). Therefore, two lattices of Cu_3Ge (0 1 0) plane match with one lattice of the c-sapphire (0 0 1) plane, as illustrated in **Figure 6(b)**. The lattice misfits of the width and length between two systems are ~4.60% and ~-2.33%, respectively. The interface region close to that of **Figure 6(a)** is enlarged for observation using HRTEM and shown in **Figure 6(c)**, to study the atomic structure in detail. The crystal planes of the film and the substrate connect with each other, but slightly bend around the misfit dislocations. Domain matching epitaxy (DME) [28–32] can be used to explain the matching of two phases, since no pseudomorphic Cu_3Ge has grown across the interface and the misfit strain is accommodated within several atomic layers at the interface. 9/8 and 8/7 domains alternate with a relative frequency of 0.5 to accommodate the misfit strain perpendicular to the interface, as shown in **Figure 6(c)**.

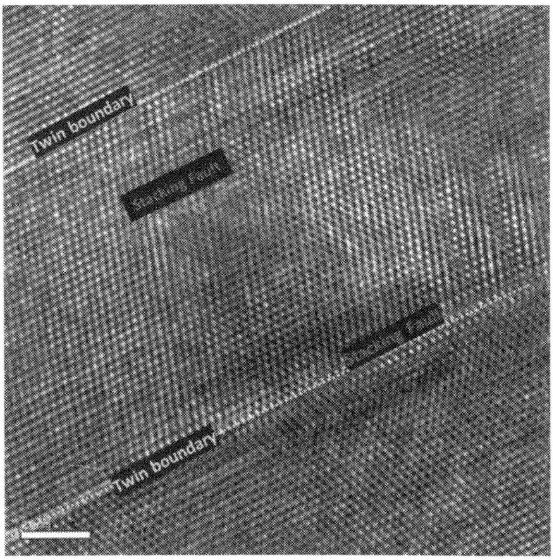

Figure 7 A typical HRTEM image showing the planar defects in Cu_3Ge thin film, in which twins and stacking faults coexist [42].

Planar defects, including twins and stacking faults (SF), are observed to exist in the as-grown epitaxial Cu_3Ge film and shown by HRTEM image (**Figure 7**). Extrinsic stacking fault is observed to exist close to the lower twin boundary (TB). Individual SF is observed within one grain. The densities of twins and SFs in the as-deposited Cu_3Ge film are close to that of pure Cu thin films with internal strain [33]. Previously in PLD-deposited films, deformation twins have been shown to be dominant [34]. Therefore, the concept of "generalized planar fault energies (GPFEs)" can explain the formation mechanism of the twins and SFs in the as-deposited Cu_3Ge film. In this concept, multiple intrinsic material properties, including stacking fault energy (SFE), unstable stacking fault energy (USFE), and unstable twin fault energy (UTE), play together to affect the twinning possibility [35]. Alloying elements have been reported to lower the SFE of metals [36], and Cu alloy has been chosen to study twin interaction phenomena in previous studies [33] for the same reason. Thus, a leading partial Cu_3Ge is more likely to nucleate and slip due to a small energy barrier. Further, the combined effect of SFE, USFE, UTE, as well as the magnitude and orientation of the local shear stress determines whether the SF is annihilated, maintained, or transformed into a twin. If the nucleation and gliding barrier for the trailing partial is low, it will be easy to annihilate an SF. Due to the existence of SF in our films, the difference between USFE and SFE should be high for Cu_3Ge since it determines the energy barrier for the trailing partial to move. Therefore, stacking faults are generated by the movement of leading partials without trailing partials. Twins are then easily generated once a leading partial is emitted and an SF is formed, since the UTE is not quite higher than the USFE.

Grain size is another parameter to affect deformation twinning in nanocrystalline metals and alloys, in addition to GPFE curves. Deformation twinning is easiest at an optimum grain size [37]. Twins in metallic thin films have been reported to lead to excellent properties, including high electromigration resistance [35], high strength and ductility, good mechanical stability, and low electrical resistivity. They can also accommodate residual strains [38]. Furthermore, Cu_3Ge with twins will be a better metallization material as the twin boundaries will deter the diffusion.

The local work function of epitaxial Cu_3Ge thin film ϕ_{Cu3Ge} was measured by KPFM, showing the fundamental electronic property which affects both electron emission through the surface and electronic trajectories near the surface [39]. KPFM measures the work function of solid surfaces at atomic or molecular scales, demonstrating information about composition and electronic state of the local structures on the surface. When the conducting tip and the sample are brought in contact, a net electric current would flow between them until the Fermi levels are aligned. During measurement, a voltage is applied between tip and sample, consisting of a DC-bias and an AC-voltage:

$$V = (V_{DC} - V_{CPD}) + V_{AC}\, sin(\omega t) \tag{1}$$

where V_{CPD} is the contact potential difference between a conductive tip and a sample, $V_{AC}sin(\omega t)$ and V_{DC} are the applied AC voltages of frequency ω and DC voltage on the tip. The electrostatic force in a capacitor can be written as:

$$F = \frac{1}{2}\frac{dC}{dz}V^2 \tag{2}$$

where C is the tip-sample capacitance, z is the tip-sample distance, and V is the tip-sample voltage. Combining the above two equations, the electrostatic force can be split up into three contributions:

$$F = F_{DC} + F_{\omega} + F_{2\omega} \tag{3}$$

The DC component (F_{DC}) contributes to the topographical signal, the term F_{ω} measures the contact potential, and the contribution $F_{2\omega}$ can be used for capacitance microscopy. They can be expressed as following:

$$F_{dc} = \frac{dC}{dz}\left[\frac{1}{2}\left(V_{DC} - V_{CPD}\right)^2 + \frac{1}{4}V_{AC}^2\right] \tag{4}$$

$$F_{\omega} = \frac{-dC}{dz}\left(V_{DC} - V_{CPD}\right)V_{AC}\, sin\left(\omega t\right) \tag{5}$$

$$F_{2\omega} = \frac{-1}{4}\frac{dC}{dz}V_{AC}^2\, cos\left(2\omega t\right) \tag{6}$$

When $V_{DC} = V_{CPD}$, the electrostatic force component measured at frequency ω and the oscillating amplitude would be zero. Consequently, V_{DC} can track V_{CPD} at each point of the scan area using a feedback circuit. Once V_{CPD} is obtained, the local work function of Cu_3Ge thin film ϕ_{Cu3Ge} can be calculated as:

$$eV_{CPD} = \varphi_{tip} - \varphi_{Cu_3Ge} \tag{7}$$

where ϕ_{tip} is the work function of the conductive tip (~4.90 + 0.02 eV) [40], calibrated with a freshly cleaved highly oriented pyrolytic graphite (HOPG, Grade 1 SPI, 4.65 eV) [40] before the experiment. **Figure 8(a)** and **(b)** shows the topographical image and the corresponding surface potential image obtained for epitaxial Cu_3Ge thin film. Small grains of tens of nanometers can be seen clearly from the topographical image, while an almost uniform distribution of V_{CPD} within each grain was observed from the contact potential image. The "band profiles" derived from the middle regions (between two white dashed lines) of **Figure 8(a)** and **(b)** are shown in **Figure 8(c)** and **(d)**, demonstrating the height and V_{CPD} variation. The average V_{CPD} for epitaxial Cu_3Ge thin film was measured to be ~0.43 V. Therefore, the work function of Cu_3Ge thin film is:

$$\varphi_{Cu_3Ge} = \varphi_{tip} - eV_{CPD} = \left(4.9 \pm 0.02\right) - 0.43 = \left(4.47 \pm 0.02\right)eV \tag{8}$$

which is between the work functions of n^+ and p^+-polysilicon [41]. This value is desirable for epitaxial Cu_3Ge thin film to be used as a mid-gap gate metal even at very low temperatures for applications in CMOS devices, because it would require minimal and symmetric channel implants even at linewidths below 0.5 μm [41].

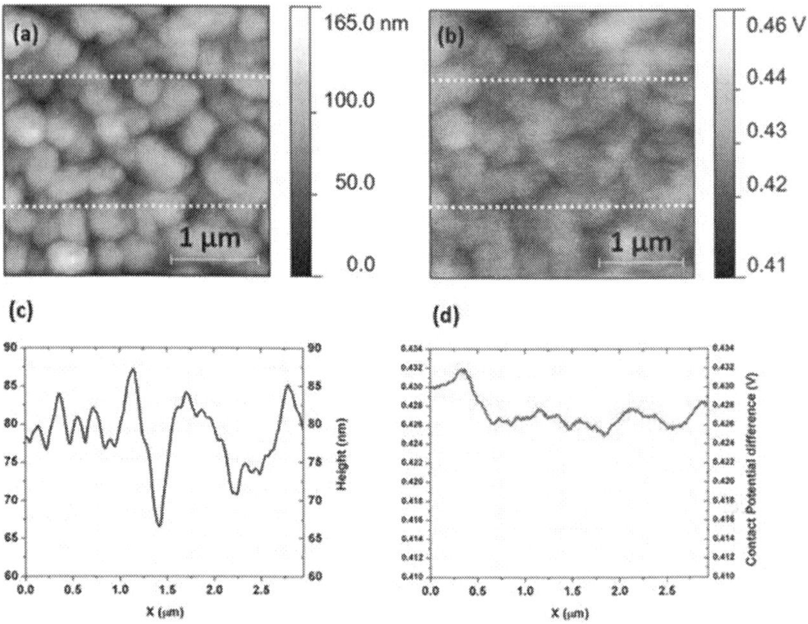

Figure 8. KPFM characterization of epitaxial Cu_3Ge thin film. (a) The topographical image, (b) the corresponding surface potential image, (c) the band profile derived from the middle region (between two white dashed lines) of (a) and (d) the band profile derived from the middle region (between two white dashed lines) of (b).

4. Conclusions

Epitaxial e_1-Cu_3Ge thin films are fabricated on *c*-plane sapphire substrate. The crystallinity of the as-grown e_1-Cu_3Ge thin film is improved, due to the modified deposition route. The e_1-Cu_3Ge phase (orthorhombic) is grown by the paradigm of domain matching epitaxy on *c*-plane sapphire, where two structures and their symmetries are completely different. In addition, twins are observed and studied in Cu_3Ge thin films, and their formation mechanism is explained by the concept of generalized planar fault energies (GPFEs). Twinning will reduce the diffusion of Cu and Ge atoms into the substrate or adjacent layers, and enhance the mechanical stability of Cu_3Ge thin films due to the improved strength and ductility and higher electromigration resistance. The average work function of epitaxial Cu_3Ge thin film is measured to be ~4.47 + 0.02 eV, rendering it a desirable mid-gap gate metal to be used for applications in CMOS devices. Therefore, the present epitaxial Cu_3Ge thin film with controlled crystallinity and defect structure is a promising candidate for the next-generation metallization material in the semiconductor industry.

http://dx.doi.org/10.5772/64060

Acknowledgements

FW and NY acknowledge the partial support by the National Science Foundation-MRSEC program through the Princeton Center for Complex Materials (DMR-0819860).

Author details

Fan Wu* and Nan Yao*

*Address all correspondence to: fanwu@princeton.edu and nyao@princeton.edu

Princeton Institute for the Science and Technology of Materials (PRISM), Princeton University, Princeton, New Jersey, USA

References

[1] Krusin-Elbaum L, Aboelfotoh MO. Unusually low resistivity of copper germanide thin films formed at low temperatures. Applied Physics Letters. 1991;58:1341–3.

[2] Doyle JP, Svensson BG, Aboelfotoh MO. Copper germanide Schottky barrier contacts to silicon. Journal of Applied Physics. 1996;80:2530–2.

[3] Huang JS, Huang SS, Tu KN, Deng F, Lau SS, Cheng SL, et al. Kinetics of Cu_3Ge formation and reaction with Al. Journal of Applied Physics. 1997;82:644–9.

[4] Borek MA, Oktyabrsky S, Aboelfotoh MO, Narayan J. Low resistivity copper germanide on (1 0 0) Si for contacts and interconnections. Applied Physics Letters. MRS Proceedings, Volume 514, 1996;69:3560–2.

[5] Aboelfotoh MO, Borek MA, Narayan J. Ohmic contact to p-type GaAs using Cu_3Ge. Applied Physics Letters. 1999;75:3953–5.

[6] Aboelfotoh MO, Oktyabrsky S, Narayan J, Woodall JM. Microstructure characterization of Cu_3Ge/n-type GaAs ohmic contacts. Journal of Applied Physics. 1994;76:5760–3.

[7] Aboelfotoh MO, Lin CL, Woodall JM. Novel low-resistance ohmic contact to n-type GaAs using Cu_3Ge. Applied Physics Letters. 1994;65:3245–7.

[8] Borek MA, Oktyabrsky S, Aboelfotoh MO, Narayan J. Properties of Cu_3Ge films for contacts to Si and SiGe and Cu metallization. In: Murarka SP, Eizenberg M, Fraser DB, Madar R, Tung R, editors. Advanced Interconnects and Contact Materials and Processes for Future Integrated Circuits; 1998. pp. 269–74.

[9] Aboelfotoh MO, Borek MA, Narayan J. Interaction of Cu and Cu_3Ge thin films with $Si_{1-x}Ge_x$ alloys. Applied Physics Letters. 1999;75:1739–41.

[10] Aboelfotoh MO, Tu KN, Nava F, Michelini M. Electrical transport properties of Cu_3Ge thin films. Journal of Applied Physics. 1994;75:1616–9.

[11] Guizzetti G, Marabelli F, Pellegrino P, Sassella A, Aboelfotoh MO. Optical response of Cu_3Ge thin films. Journal of Applied Physics. 1996;79:8115–7.

[12] Kumar D, Vispute RD, Aboelfotoh O, Oktyabrsky S, Jagannadham K, Narayan J, et al. $LaNiO_3$ and Cu_3Ge contacts to $YBa_2Cu_3O_{7-x}$ films. Journal of Electronic Materials. 1996;25:1760–6.

[13] Aboelfotoh MO, Borek MA, Narayan J. Microstructure and electrical resistivity of Cu and Cu_3Ge thin films on $Si_{1-x}Ge_x$ alloy layers. Journal of Applied Physics. 2000;87:365–8.

[14] Peter AP, Carbonell L, Schaekers M, Adelmann C, Meersschaut J, Franquet A, et al. Selective chemical vapor synthesis of Cu_3Ge: process optimization and film properties. Intermetallics. 2013;34:35–42.

[15] Hsin H-C, Lin W-T, Gong JR, Fang YK. Cu_3Ge Schottky contacts on n-GaN. Journal of Materials Science: Materials in Electronics. 2002;13:203–6.

[16] Liang HH, Luo JS, Lin WT. Room temperature oxidation of Cu_3Ge and $Cu^{-3}(Si_{1-x}Ge_x)$ on $Si_{1-x}Ge_x$. Materials Science in Semiconductor Processing. 2001;4:233–5.

[17] Nath P, Chopra KL. Electrical resistivity and thermoelectric power of copper germanium films. Thin Solid Films. 1979;58:339–43.

[18] Rao SS, Prater JT, Wu F, Nori S, Kumar D, Yue L, et al. Positive exchange bias in epitaxial permalloy/MgO integrated with Si (1 0 0). Current Opinion in Solid State and Materials Science. 2014;18:140–6.

[19] Rao SS, Prater J, Wu F, Shelton C, Maria J-P, Narayan J. Interface magnetism in epitaxial $BiFeO_3$-$La_{0.7}Sr_{0.3}MnO_3$ heterostructures integrated on Si (1 0 0). Nano Letters. 2013;13:5814–21.

[20] Lee Y, Wu F, Kumar R, Hunte F, Schwartz J, Narayan J. Epitaxial integration of dilute magnetic semiconductor Sr3SnO with Si (0 0 1). Applied Physics Letters. 2013;103:112101.

[21] Bayati M, Molaei R, Wu F, Budai J, Liu Y, Narayan R, et al. Correlation between structure and semiconductor-to-metal transition characteristics of VO_2TiO_2/sapphire thin film heterostructures. Acta Materialia. 2013;61:7805–15.

[22] Molaei R, Bayati R, Wu F, Narayan J. A microstructural approach toward the effect of thickness on semiconductor-to-metal transition characteristics of VO_2 epilayers. Journal of Applied Physics. 2014;115:164311.

[23] Lee Y, Wu F, Narayan J, Schwartz J. Oxygen vacancy enhanced room-temperature ferromagnetism in Sr3SnO/c-YSZ/Si (0 0 1) heterostructures. MRS Communications. 2014;4:7–13.

[24] Rao S, Prater J, Wu F, Nori S, Kumar D, Narayan J. Integration of epitaxial permalloy on Si (1 0 0) through domain matching epitaxy paradigm. Current Opinion in Solid State and Materials Science. 2013;18:1–5.

[25] Singamaneni SR, Prater J, Wu F, Narayan J. Interface magnetism of two functional epitaxial ferromagnetic oxides integrated with Si (1 0 0). Bulletin of the American Physical Society. APS March Meeting Abstracts, volume 1, page 1249, 2014.

[26] Singamaneni SR, Prater J, Wu F, Nori S, Kumar D, Yue L, et al. Positive exchange bias in epitaxial permalloy/MgO integrated with Si (1 0 0). Bulletin of the American Physical Society. APS March Meeting Abstracts, volume 1, page 1254, 2014.

[27] Wu F, Narayan J. Controlled epitaxial growth of body-centered cubic and face-centered cubic Cu on MgO for integration on Si. Crystal Growth & Design. 2013;13:5018–24.

[28] Narayan J, Larson BC. Domain epitaxy: a unified paradigm for thin film growth. Journal of Applied Physics. 2003;93:278–85.

[29] Wu F, Rao SS, Prater JT, Zhu YT, Narayan J. Tuning exchange bias in epitaxial Ni/MgO/TiN heterostructures integrated on Si(1 0 0). Current Opinion in Solid State and Materials Science. 2014;18:263–8.

[30] Singamaneni SR, Fan W, Prater JT, Narayan J. Magnetic properties of $BaTiO_3$/$La_{0.7}Sr_{0.3}MnO_3$ thin films integrated on Si(1 0 0). Journal of Applied Physics. Issue 22, pages 224–104, 2014;116.

[31] Bayati R, Molaei R, Richmond A, Nori S, Wu F, Kumar D, et al. Modification of properties of Yttria stabilized Zirconia epitaxial thin films by Excimer laser annealing. ACS Applied Materials & Interfaces. 2014;6:22316–25.

[32] Gbordzoe S, Kotoka R, Craven E, Kumar D, Wu F, Narayan J. Effect of substrate temperature on the microstructural properties of titanium nitride nanowires grown by pulsed laser deposition. Journal of Applied Physics. 2014;116:4310.

[33] Wu F, Wen HM, Lavernia EJ, Narayan J, Zhu YT. Twin intersection mechanisms in nanocrystalline fcc metals. Materials Science and Engineering: A. 2013;585:292–6.

[34] Wu F, Zhu YT, Narayan J. Grain size effect on twin density in as-deposited nanocrystalline Cu film. Philosophical Magazine. 2013;93:4355–63.

[35] Zhu YT, Liao XZ, Wu XL. Deformation twinning in nanocrystalline materials. Progress in Materials Science. 2012;57:1–62.

[36] Venables JA. The electron microscopy of deformation twinning. Journal of Physics and Chemistry of Solids. 1964;25:685–92.

[37] Wu XL, Zhu YT. Inverse grain-size effect on twinning in nanocrystalline Ni. Physical Review Letters. 2008;101:025503.

[38] Wu F, Zhu YT, Narayan J. Macroscopic twinning strain in nanocrystalline Cu. Materials Research Letters. 2014;2:63–9.

[39] Fain SC, McDavid JM. Work-function variation with alloy composition: Ag–Au. Physical Review B. 1974;9:5099–107.

[40] Yan F, Schoofs F, Shi J, Ha SD, Jaramillo R, Ramanathan S. Local charge writing in epitaxial $SmNiO_3$ thin films. Journal of Materials Chemistry C. 2014;2:3805–11.

[41] Aboelfotoh MO, Krusin-Elbaum L, Sun YC. Wet etching process with high selectivity between Cu and Cu_3Ge. Google Patents Publication number EP0769808 A3. 1997.

[42] Wu F, Zheng JK, Cai W, Yao N, Zhu YT, Narayan J. Fabrication of epitaxial Cu_3Ge on sapphire with controlled crystallinity and planar defects. Journal of Alloys and Compounds. 2015;641:238–43.

Chapter 5

Nanomachining of Fused Quartz Using Atomic Force Microscope

Yoshio Ichida

Additional information is available at the end of the chapter

http://dx.doi.org/10.5772/64408

Abstract

Nanomachining experiments on fused quartz surface have been performed using an atomic force microscope combined with a two-axis capacitive force/displacement transducer. The minimum normal force f_{nP} needed to form reproducibly a groove was about 4.7 μN. The minimum critical normal force f_{nR}, tangential force f_{tR}, and groove depth d_{gR} when the material removal process began were found to be 33.7 μN, 18.7 μN, and 4.3 nm, respectively. Characteristic changes in the swelling ratio R_s and the ratio of force components at the critical normal force f_{nR} can be used to identify the critical condition for changing from plastic deformation to material removal process region.

Keywords: nanomachining, atomic force microscope, material removal process, fused quartz, three-sided pyramidal diamond tip

1. Introduction

Recently, surface modifications with a scanning probe microscope have attracted particular interest as a nanomachining technique for processing surfaces on a nanometer scale [1–4]. Especially, nanomachining methods using an atomic force microscope (AFM) will play an increasingly important role in an ultrafine fabrication technology to produce various Nano/microdevices regardless of conductivity [5–18]. However, most of the previous studies on nanomachining have been carried out using a cantilever-type AFM [19–25]. In this AFM, the probe tip as a machining tool is asymmetrically supported by the cantilever. Therefore, because of the asymmetric deflection of the cantilever, the tool tip cannot contact the work surface symmetrically and accurately. To analyze the precise mechanism of nanometer-scale machining, it is not suitable to use such a cantilever-type AFM.

The objective of this chapter is to clarify the fundamental mechanisms of material removal in nanomachining using an AFM diamond tip. To investigate the critical conditions needed to initiate the material removal process, such as the minimum normal force, tangential force, and groove depth when the removal process begins, a series of nanomachining experiments on polished quartz surfaces was conducted using an AFM combined with a two-axis capacitive force/displacement transducer [26–30].

2. Experimental procedure for nanomachining

Figure 1 shows the experimental method for nanomachining used in this study and its cross-sectional machining model is illustrated in **Figure 2**. By means of the AFM diamond tip as a machining tool, the straight nanometer-scale grooves were machined under a constant normal machining force f_n and a constant machining speed v_m. The tangential machining force f_t was measured experimentally and was used to elucidate the material removal mechanism in nanomachining.

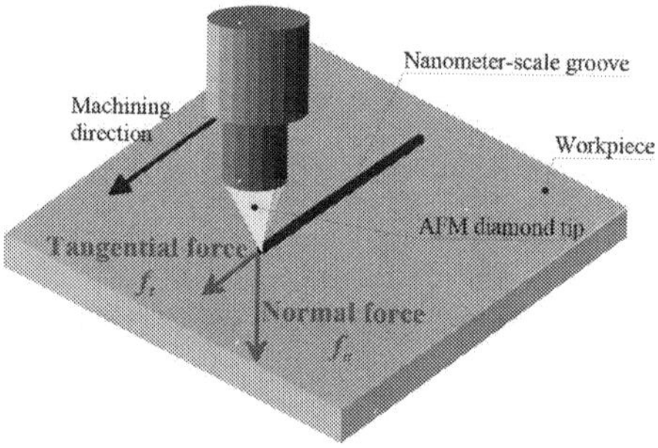

Figure 1. Experimental method for constant-force nanomaching.

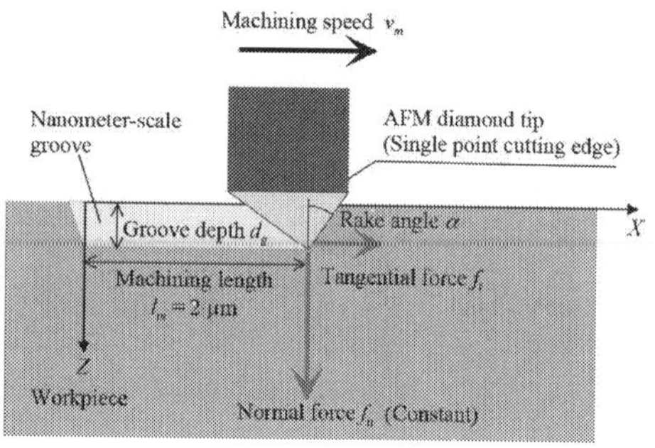

Figure 2. Cross-sectional machining model for constant-force nanomachining with AFM diamond tip.

Nanomachining experiments on fused quartz were performed using an AFM combined with a two-axis capacitive force/displacement transducer, a schematic illustration of which is shown in **Figure 3**. The *x*-axis transducer (lateral force transducer) is comprised of two additional sensors that are mounted transversely to the *z*-axis force transducer (normal force transducer). Using this two-axis transducer, the force and the displacement in both the *z*- and *x*-axes were measured.

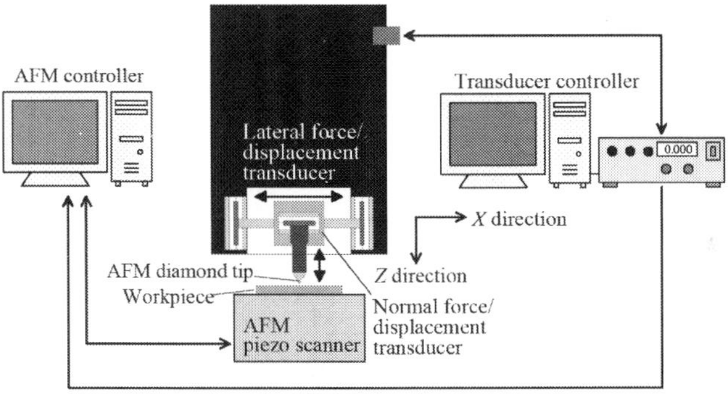

Figure 3. Schematic illustration of constant-force nanomachining equipment with AFM.

A three-sided pyramidal diamond tip (cube corner tip) with a radius of 50 nm set symmetrically in the center of the *z*-axis force transducer was used for both nanomachining and imaging in these experiments. The shape of the diamond tip, the SEM image of its top edge, and the machining direction used for processing straight grooves are shown in **Figure 4**. Experimental conditions for processing straight grooves are shown in **Table 1**. After the nanomachining experiments, the modified surface was immediately imaged at a normal force of about 0.5 μN using the same diamond tip.

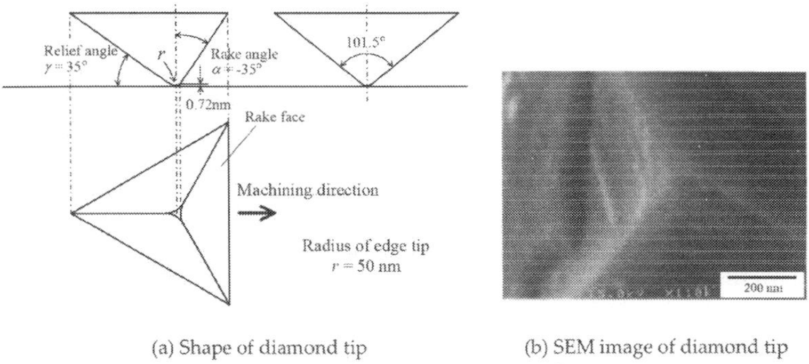

Figure 4. Dimensions of AFM tip and its SEM image. (a) Shape of diamond tip. (b) SEM image of diamond tip.

Machining tool	Three-sided pyramidal diamond tip Tip edge radius: 50 nm Rake angle: –35° Relief angle: 35°
Machining length l_m	2 μm
Machining speed v_m	100 nm/s
Nominal normal force F_n	1, 2, 4, 6, 8, 10, 20, 30, 40, 50, 60, 70, 80, 90, 100 μN
Work material	Fused quartz Nanohardness: 8.7 GPa Elastic modulus: 72 GPa Surface roughness: 0.5 nm*Ra*
Atmosphere	In air Temperature: 22°C Humidity: 50%

Table 1. Machining conditions.

3. Formation of nanometer-scale grooves

Typical AFM images of the straight grooves with a length of 2 μm processed at different normal forces f_n of 48.2 μN or less are shown in **Figure 5**. In this figure, the actual normal machining force f_n measured with the *z*-axis force transducer, the A-A′ sectional profiles of grooves, and the actual mean groove depths d_g obtained from these profiles are also indicated. Grooves were not formed using normal machining forces of 3.8 μN or less.

But, at a normal force of 5.6 μN a very fine straight groove with a constant depth of about 0.26 nm and a constant width of about 20 nm was successfully generated. From this result, it is confirmed that the minimum normal force needed to form reproducibly a groove on the fused quartz surface is between 3.8 and 5.6 μN. This minimum force is a critical value (denoted by f_{nP}) when the plastic deformation begins in this nanomachining. The value of f_{nP} is given by 4.7 μN that is the average of 5.6 and 3.8 μN.

Both depth and width of the straight groove are increased with increasing the normal force. At the same time, small amounts of residual debris can be observed at the end of the groove when the normal force exceeded 28.7μN. **Figure 6** shows a typical 3D AFM image of the groves and the residual debris and **Figure 7(a)** and **(b)** show the sectional profiles at sections AA′ and BB′ in **Figure 6**, respectively. The bottom of the straight groove has a smooth surface with a roughness less than a few nanometers *Rz*. The side swelling (swell-out residual) is observed at both sides of the straight groove. On the other hand, the protrusion height of the residual debris is much higher than that of the side swelling. Therefore, this residual debris seems to offer evidence that material removal action has occurred.

http://dx.doi.org/10.5772/64408

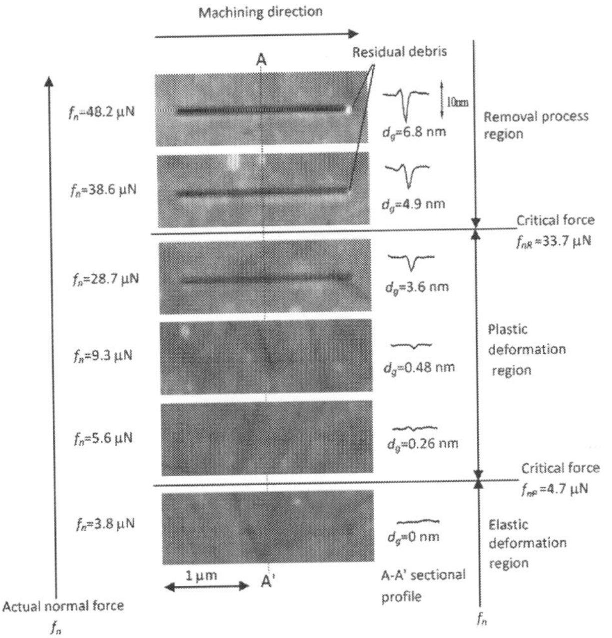

Figure 5. Typical AFM images of the straight grooves with a length of 2 μm processed at various normal forces f_n less than 48.2 μN.

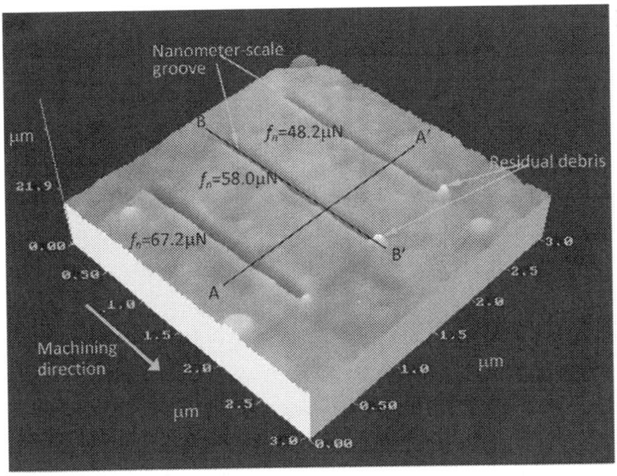

Figure 6. AFM image of machined nanometer-scale grooves.

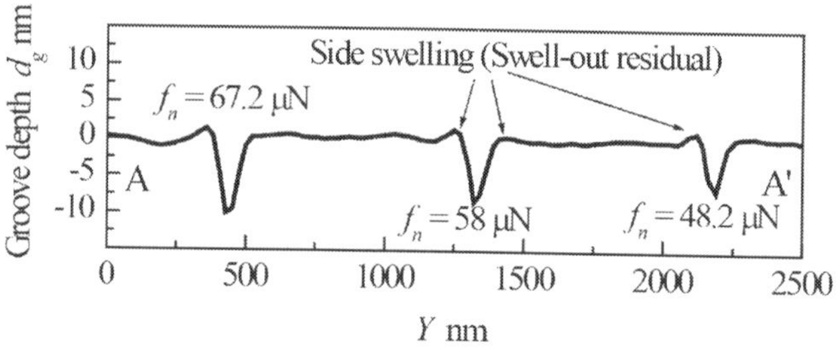

(a) Profile at section AA' in Fig. 6

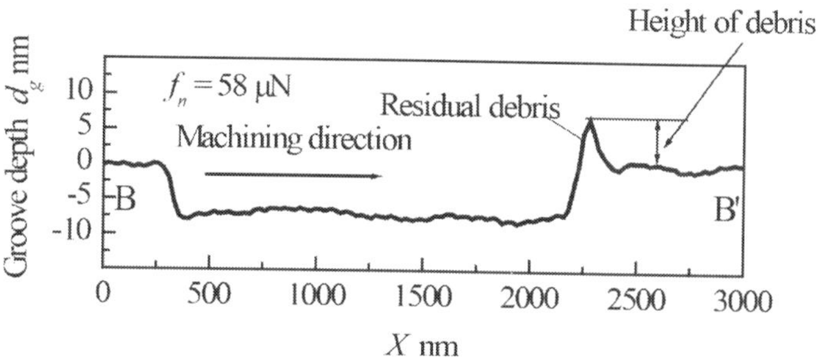

(b) Profile at section BB' in Fig. 6

Figure 7. Section profiles of machined nanometer-scale grooves. (a) Profile at section AA′ in **Figure 6**. (b) Profile at section BB′ in **Figure 6**.

Especially, when the AFM probe scanned the groove surface, the cutting chip that remained at the end of the groove was fractured by the probe. Consequently, such residual debris remained at the end of the groove. Generally, it is easy for the residual cutting chip to be fractured by the probe in scanning for imaging because it projects from the surface. Hence, it is very difficult to observe the cutting chips that remained at the end of the groove after nanomachining, using the AFM.

So we tried observing the cutting chips with a field emission scanning electron microscope (FE-SEM/SU8040). **Figure 8** shows the typical observation results. When the normal force was less than 28.7 μN, no cutting chips were observed. But when the normal force was 38.6 μN, a nanometer-scale cutting chip that remained at the end of the groove was grasped. These results show that the minimum normal force when the chip begins to be produced on the fused quartz surface is between 28.7 and 38.6 μN.

http://dx.doi.org/10.5772/64408

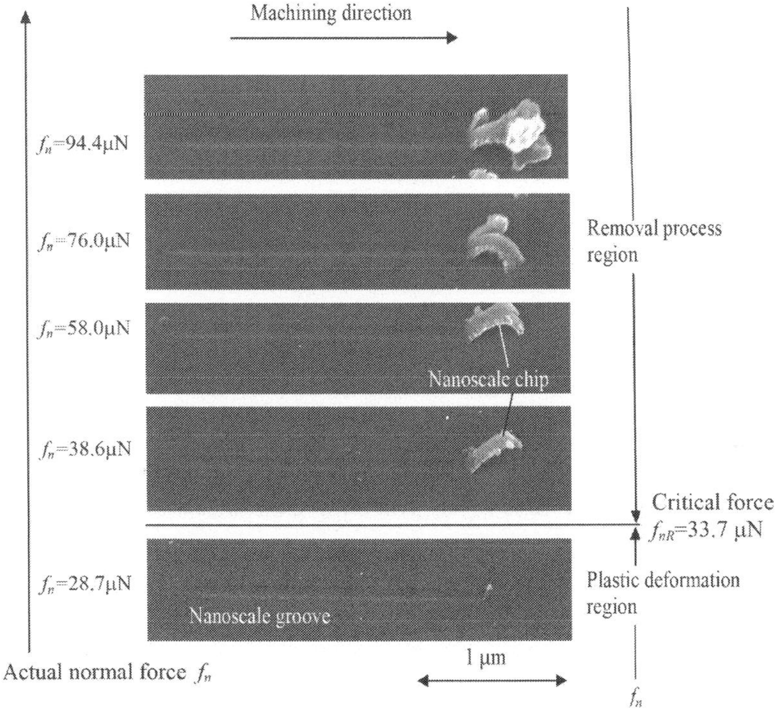

Figure 8. Typical AFM images of the straight grooves with a length of 2 μm processed at various normal forces f_n of 94.4 μN or less.

Figure 9 shows the high magnification SEM images of the cutting chips obtained at the normal forces of 38.6, 48.2, and 94.4 μN. Because these cutting chips have a continuous shape and a smooth surface, they can be judged to be a flow-type chip.

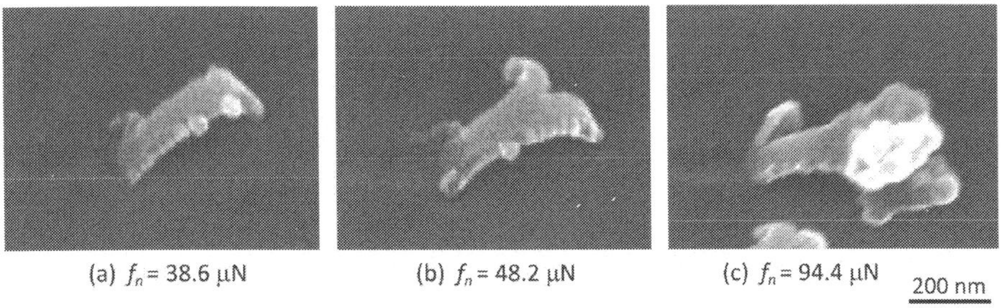

Figure 9. High magnification SEM images of cutting chips. (a) f_n = 38.6 μN. (b) f_n = 48.2 μN. (c) f_n = 94.4 μN.

These results show that the mechanism of the groove formation, that is, the material removal in nanomachining, is based on the ductile mode minute cutting action that is chiefly performed by the shear deformation. The size of the cutting chip increases with increasing normal force.

These AFM and SEM observations show that the minimum groove depth when the material removal process begins is between 3.6 and 4.9 nm. This minimum groove depth d_{gR} is an important critical value to indicate the machining unit in this nanomachining process [31], and is given by 4.3 nm as an average of the upper and lower bound values (4.9 and 3.6 nm, respectively). Simultaneously, it is found that the minimum normal force to initiate material removal is between 28.7 and 38.6 μN. The minimum force f_{nR} is also an important critical value to indicate the machining unit and is given by 33.7 μN (the average of the upper bound value of 38.6 and the lower bound value of 28.7 μN).

4. Transient behavior from plastic deformation to removal process

The effects of the normal force on the characteristic machining parameters that indicate the features of the groove processed by the diamond edge, that is, the groove depth d_g, the groove width w_g, the aspect ratio d_g/w_g, and the swelling ratio R_s, are shown in **Figure 10**. The swelling ratio R_s is given by $R_s = (B_1 + B_2)/A$, where B_1 and B_2 are cross-sectional areas of the swill-out residuals, and A is a cross-sectional area of the groove, respectively, as shown by the small illustration in **Figure 10**.

From the above-mentioned results, it is found that the following three regions exist in the process until the fused quartz surface is removed by the diamond tip. These regions are:

(a) $0 \leq f_n \leq 4.7$ μN: Elastic deformation region

(b) $4.7 \leq f_n \leq 33.7$ μN: Elastic/plastic deformation region

(c) $33.7 \leq f_n$ ($d_g \geq 4.3$ nm): Removal process (cutting process) region

In these expressions, 4.7 μN is the critical normal force f_{nP} when plastic deformation originates, and its standard deviation is 0.73 μN. On the other hand, 33.7 μN and 4.3 nm are the critical normal force f_{nR} and groove depth d_{gR} when removal action originates, and these standard deviations are 4.1 μN and 0.62 nm, respectively.

As shown in **Figure 10**, the swelling ratio R_s decreases rapidly when the normal force exceeds its critical value f_{nR}. This feature change in R_s can be used to identify a transition condition for changing from the plastic deformation to the removal process region.

The changes of tangential force f_t and ratio of force components with increasing normal force f_n are shown in **Figure 11**. These results show that the minimum tangential force, namely, the critical value f_{tP} when the plastic deformation begins, is given by 2.2 μN as an average of the upper and lower bounds (2.46 and 1.96 μN, respectively). Concurrently, the mean minimum tangential force f_{tR} when the material removal process begins is 18.7 μN (the average of the upper bound/22.8 μN and lower bound/14.5 μN).

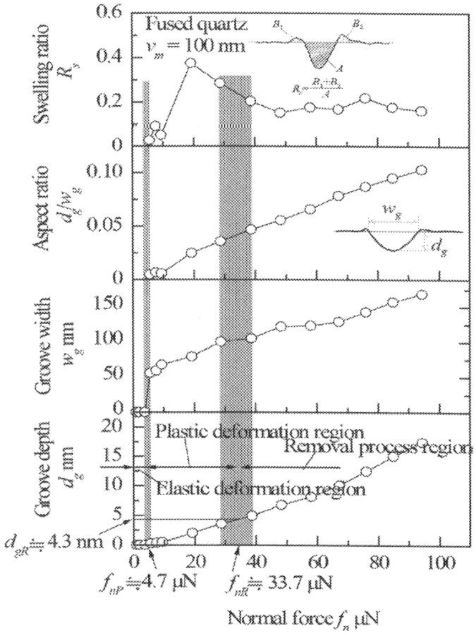

Figure 10. Changes of characteristic parameters with increasing normal force f_n.

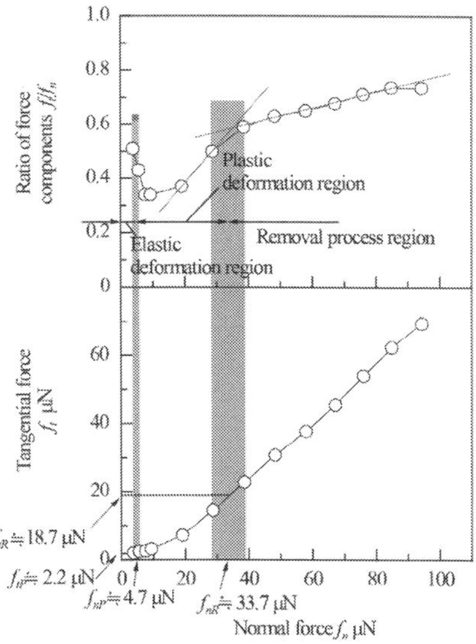

Figure 11. Effects of normal force on tangential force and ratio of force components.

The main peculiarity in **Figure 11** is a change in the ratio of the force components with increasing the normal force f_n. Especially, the value of f_t/f_n changes suddenly from a rapid increase to a gradual increase with increasing the normal force. This feature change in f_t/f_n is caused by the difference in the direction of the processing force between the plastic deformation and removal process regions [17, 26]. This phenomenon can also be used to identify a critical condition when changing from the plastic deformation to the removal process region.

5. Conclusions

Nanometer-scale machining on fused quartz surfaces has been carried out with an AFM combined with a two-axis capacitive force/displacement transducer. The main results obtained in this study are:

(1) The minimum normal force f_{nP} required to form reproducibly a groove is about 4.7μN. Namely, the groove is first formed by plastic deformation when the normal force exceeds 4.7 μN.

(2) The minimum critical normal force f_{nR}, tangential force f_{tR}, and groove depth d_{gR} required to initiate the material removal process are, respectively, as follows:

$$f_{nR} = 33.7\,\mu N$$

$$f_{tR} = 18.7\,\mu N$$

$$d_{gR} = 4.3\,nm$$

Namely, the groove is first formed by cutting action when the groove depth exceeds 4.3 nm.

(3) The characteristic changes in the swelling ratio R_s and the ratio of force components f_t/f_n at the critical normal force f_{nR} can be used to identify a critical condition for changing from the plastic deformation to the material removal process region.

(4) The mechanism of the material removal in nanomachining is based on the ductile mode minute cutting action that is chiefly performed by the shear deformation.

As described above, the minimum groove depth, that is, the minimum depth of cut when the groove is first removed by the cutting action is a fundamental value called "Machining Unit," which is an important parameter that controls machining accuracy and machining quality [31]. This machining unit will be mainly affected by the tip radius of the diamond tool. To reduce the machining unit, it is necessary to reduce the tip radius of the diamond tool. In the future, we are planning to report on the effects of the tip radius on the machining unit.

http://dx.doi.org/10.5772/64408

Acknowledgements

This research was supported in part by Grants-in-Aid for General Science Research (C) (No. 19560106) from the Ministry of Education, Culture, Sports, Science and Technology of Japan.

Author details

Yoshio Ichida

Address all correspondence to: ichida@cc.utsunomiya-u.ac.jp

CBN & Diamond Nanomachining Institute, Utsunomiya, Japan

References

[1] Snow E S, Campbell, P M: Fabrication of Si nanostructures with an atomic force microscope, Applied Physics Letters, 1993, 64, 15, 1932–1934. ISSN: 0003-6951

[2] Tsau K, Wang D, Wang K L: Nanometer scale patterning of silicon (100) surfaces by an atomic force microscope operating in air, Applied Physics Letters, 1994, 64, 16, 2133–2135. ISSN: 0003-6951

[3] Wendel M, Kühn S, Lorenz H, Kotthaus J P: Nanolithography with an atomic force microscope for integrated fabrication of quantum electronic devices, Applied Physics Letters, 1994, 65, 14, 1775–1777. ISSN: 0003-6951

[4] Bhushan B, Koinkar V N K: Nanoindentation hardness measurements using atomic force microscopy, Applied Physics, 1994, 64, 13, 1653–1655. ISSN: 0021-8979

[5] Samitsu Y: A study of atomic scale processing using an atomic force microscope, Journal of Japan Society Precision Engineering, 1995, 61, 8, 1121–1125. ISSN0912-0289

[6] Bouchiat V, Esteve D: Lift off lithography using an atomic force microscope, Applied Physics Letters, 1996, 69, 3098–3100. ISSN: 0003-6951

[7] Tegen S, Kracke B, Damaschke B: Surface modifications with a scanning force microscope, Review of Scientific Instruments, 63, 1458–1464. ISSN: 0034-6748

[8] Li S F Y, Ng H T P, Zhang C, Ho P K H, Zhou L, Bao G W, Chan S L H: Submicrometer lithography of a silicon substrate by machining of photoresist using atomic force microscopy followed by wet chemical etching, Nanotechnology, 1997, 8, 76–81. ISSN: 0957-4484

[9] Miyazaki T, Yoshioka S, Shirai Y, Misu T, Taniguchi N: Removal characteristics of processing with SPM, Annals of the CIRP, 1998, 47, 1, 153–156. ISSN: 0007-8506

[10] Ichida Y, Sato R, Morimoto Y, Takahashi K: Nanometer-scale machining of single-crystal silicon using an atomic force microscope, JSPE Publication Series, 1999, 3, 677–682.

[11] Ichida Y, Takahashi K: Nanometer-scale scratching on the single-crystal silicon surface using an atomic force microscope, Proceedings of the 3rd International Conference on Abrasive Technology, 1999, Brisbane, Australia, 421–426.

[12] Ichida Y, Morimoto Y, Sato R: Nanoscale mechanical processing of silicon (001) surface by atomic force microscope diamond tip, Proceedings of the 1st euspen Topical Conference on Fabrication and Metrology in Nanotechnology, 2000, 1, Copenhagen, Denmark, 206–210.

[13] Hang T H, Chang J: Effects of AFM-based nanomachining process on aluminum surface, Journal of Physics and Chemistry of Solids, 2003, 64, 913–918.

[14] Chen Y J, Hsu J H, Lin N H: Fabrication of metal nanowires by atomic force microscopy nanoscratching and lift-off process, Nanotechnology, 2005, 16, 1112–1115. ISSN: 0957-4484.

[15] Ashida K, Morita N, Yoshida Y: Study on nano-machining process using mechanism of a friction force microscope, JSME International Journal (Ser.-C)-Mechanical Systems, Machine Elements and Manufacturing, 2001, 44, 244–253.

[16] Ichida Y, Sato R, Morimoto Y: Material removal characteristics in nanometer scratching of single crystal silicon using an AFM diamond tip, Proceedings of International Tribology Conference, 2001, Nagasaki, Japan, 763–767.

[17] Ichida Y, Morimoto Y, Sato R, Murakami M: Nanomachining on Si (100) surface using an atomic force microscope with a lateral force transducer, Technical Proceedings of 2003 Nanotechnolgy Conference, San Francisco, USA, 2003, 1, 534–537.

[18] Kawasegi N, Takano N, Oka D, Morita N, Yamada S, Kanda K, Takano S, Obata T, Ashida K: Nanomachining of silicon surface using atomic force microscope with diamond tip, Journal of Manufacturing Science and Engineering-Transactions of the ASME, 2006, 128, 723–729. ISSN: 1087-1357.

[19] Yan Y D, Sun T, Liang Y, Dong S: Investigation on AFM-based micro/nano-CNC machining system, International Journal of Machine Tool and Manufacture, 2007, 46, 11, 1651–1659. ISSN: 0890-6955

[20] Yan Y D, Hu Z J, Sun T, Zhao X S, Dong S: AFM-based machining technique of three dimensional curved microstructures, Proceedings of the 10th International Conference of European Society Precision Engineering & Nanotechnology, 2008, Zurich, Switzerland, 2, 400–404.

[21] Tseng A A, Shirakashi J, Nishimura S, Miyashita K, Li Z: Nanomachining of permalloy for fabricating nanoscale ferromagnetic structures using atomic force microscopy, Journal of Nanoscience and Nanotechnology, 2010, 10, 1, 456–466. ISSN: 1533-4880

[22] Tseng A A: Three-dimensional patterning of nanostructures using atomic force microscopes, Journal of Vacuum Science & Technology B, 2011, 29, 040801. ISSN: 2166-2746

[23] Wang Z, Jiao N, Tung S, Dong Z: Atomic force microscopy-based repeated machining theory for nanochannels on silicon oxide surfaces, Applied Surface Science, 2011, 257, 3627–3631. ISSN: 0169-4332

[24] Dong Z, Wejinya U C: Atomic force microscopy based repeatable surface nanomachining for nanochannels on silicon substrates, Applied Surface Science, 2012 258, 8689–8695. ISSN: 0169-4332

[25] Lee S H: Analysis of ductile mode and brittle of AFM machining of silicon, International Journal of Machine Tool and Manufacture, 2012, 61, 71–79. ISSN: 0890-6955

[26] Ichida Y, Sato R, Morimoto Y, Sasanuma M: Basic study of nanogrinding-nanoscale removal process in constant force grinding, Journal of Japan Society for Precision Engineering, 2006, 72, 8, 988–993. ISSN: 0912-0289

[27] Ichida Y, Yamaguchi T, Sasanuma M: Mechanism of material removal in nanomachining using AFM diamond tip, Proceedings of the 5th International Conference on Leading Edge Manufacturing in 21st Century, Osaka, Japan, 2009, 315–318.

[28] Ichida Y, Yamaguchi T, Sasanuma M: Mechanism of material removal in nanomaching of fused quartz surface using AFM diamond tip, JSME Journal of Advanced Mechanical Design, Systems, and Manufacturing, 2010, 4, 5, 1015–1021. ISSN: 1881-3054

[29] Ichida Y, Iitsuka Y, Sato R: Atomic-scale processing of sputter-deposited Pt-Pd alloy film surface using AFM diamond tips, Proceedings the 10th International Conference of European Society for Precision Engineering & Nanotechnology, 2011, Lake Como, Italy, 2, 305–308.

[30] Ichida Y: Nanometer-scale machining of fused quartz surface with AFM diamond tip, Proceedings of the 6th International Conference on Micro Manufacturing, Tokyo, Japan, 2011, 621–626.

[31] Nakazawa H: Principle of Precision Engineering, Oxford University Press, 1994, 190–196.

Chapter 6

Fabrication and Characterization of Organic–Inorganic Hybrid Perovskite Devices with External Doping

Kongchao Shen, Hao Liang Sun, Gengwu Ji,
Yingguo Yang, Zheng Jiang and Fei Song

Additional information is available at the end of the chapter

http://dx.doi.org/10.5772/63771

Abstract

Owing to its excellent light harvesting, high-charge carrier mobility, and long electron- and hole-transport lengths, organic–inorganic lead halide perovskite solar cells have attracted enormous attention recently under the urgent demands of green energy with environmental friendliness. Although various photovoltaic architectures based on alkylammonum lead halides have been fabricated and have achieved impressive power conversion efficiencies (PCEs), there are still several issues that need to be further addressed and solved properly, for example, the requirement of facile fabrication procedure, the chemical stability of perovskite films, and the environmental friendliness. Herein, we review the recent experimental progress on the external doping of hybrid perovskite devices by organics and metals, which demonstrate the tuning of optical absorption gap and the enhancement of both devices' stability and performance. Doping at varying layers in the perovskite films was discovered to contribute differently to the improvement of the hybrid organic–inorganic electronics. In the end, prospective was also made on the development of hybrid organic–inorganic devices.

Keywords: hybrid organic electronics, perovskite solar cells, external doping, device performance, stability

1. Introduction

Depletion of fossil fuels and the shortage of energy has become one of the most serious problems nowadays, which strongly restricts the sustainable development in our society. To achieve a sustainable society, we need methods of converting energy from other resources, such as wind and sunlight. Among the proposed approaches, photoelectrochemical devices offer the promise of solar fuel production through artificial photosynthesis and led to considerable development in numerous areas related to photovoltaic cells and electronics. Thin-film solar cells, such as dye-sensitized solar cells, organic photovoltaics, and colloidal nanocrystal solar cells, can be assembled with low-cost materials and manufactured with cost-effective methods and are considered very promising renewable energy technologies. Especially, hybrid organic–inorganic perovskites based on the metal halides have emerged as one class of promising light-harvesting materials because of their exceptional properties such as direct band gaps, large absorption coefficients, and high carrier mobility. Meanwhile, the organic–inorganic perovskite solar cell also have many other advantages compared to the conventional solar cells, such as easy solution process, low processing cost, extremely high power conversion efficiency which reached over 20% recently. All these merits have granted perovskite a promising candidate for the next-generation solar cells.

Although great success in the use perovskite solar cells has been witnessed over the past few years, there are still several bottlenecks which limit significantly the wide deployment of its outdoor application: for example, the poor stability of the perovskite film and the charge transport layers, dangers and health risks arising from toxic and harmful element due to the usage of lead and other heavy metal atoms. Based on these urgent demands and concerns, the improvement of perovskite electronics has been explored and demonstrated. Interestingly, doping of perovskite electronics seems to be an efficient way to enhance both the stability and performance of perovskite electronics, which is at moment classified into several approaches: doping into the electron transport layer, doping into the perovskite films, and doping into the hole transport layer. In this chapter, we review the progress of in situ fabrication and characterization of organic–inorganic hybrid electronics by external doping and propose some of them.

2. Structure and performance of perovskite electronics

Since 2009, Kojima and his coworkers used $CH_3NH_3PbBr_3$ and $CH_3NH_3PbI_3$ as sensitizers in solar cell; great progress has been made for such kind of solar cells not only due to the fact that the power conversion efficiency (PCE) is increased from 3.8 to 20.8% [1,2], but also that it has high absorption characteristics, appropriate direct band gaps, high carrier mobility, long charge carrier diffusion length, low cost, and easy fabrication processes [3]. All these excellent performances are mostly originated from the **organic–inorganic** perovskite film. The formula of perovskite is usually written as ABX_3, where A is an organic cation, e.g., $CH_3NH_3^+$, B is a metal cation, e.g., Pb^{2+}, and X is a halide anion, as shown in **Figure 1a**.

http://dx.doi.org/10.5772/63771

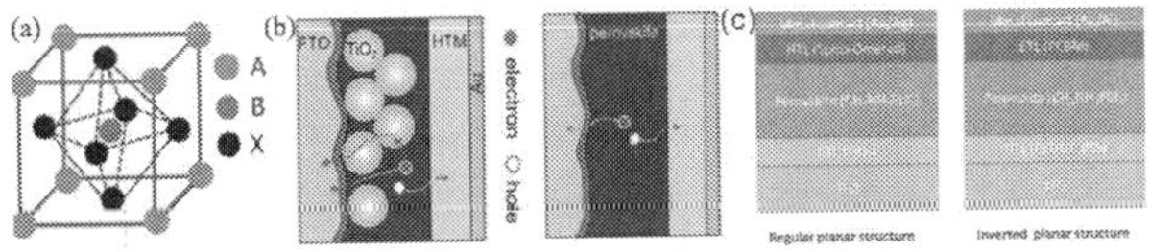

Figure 1. (a) Illustration of the organic–inorganic perovskite structure; (b) mesoporous and planar structure of hybrid perovskite devices; (c) the comparison between regular and inverted structure of hybrid perovskite device.

The sandwich-like structures of the organic–inorganic hybrid perovskite devices are shown in **Figure 1b**. Mesoporous structure and planar structure (**Figure 1b**) are the so-called positive structure, while the inverted structure is presented in **Figure 1c** with the electron transport layer (ETL) and the hole transport layer (HTL) upside down as compared to the regular perovskite structure device. As discovered, the inverted structure can reduce J-V hysteresis and is more stable due to the hydrophobic PCBM ([6,6]-phenyl-C61-butyric acid methyl ester) and its easy fabrication. With the thin perovskite layer placed between the HTL and the ETL under illumination, voltage/current is formed by generating and directionally moving the excitons (holes and electrons) after overcoming the energy barrier of the band gap. Electron–hole pairs are generated almost instantaneously after photo excitation and dissociated in the time scale of several ps, followed by the formation of high mobile charges in the neat perovskite [4]. After the generation and separation of excitions, the holes and electrons diffuse to the opposite site of the sensitizer to reach electrodes forming electrical current; however, not all of the holes and electrons can get to the electrode and contribute to the power generation, since some of them will be captured by the defects or recombined with other electrons/holes on the way to electrodes. The routes of carrier transmission are illustrated in **Figure 2**.

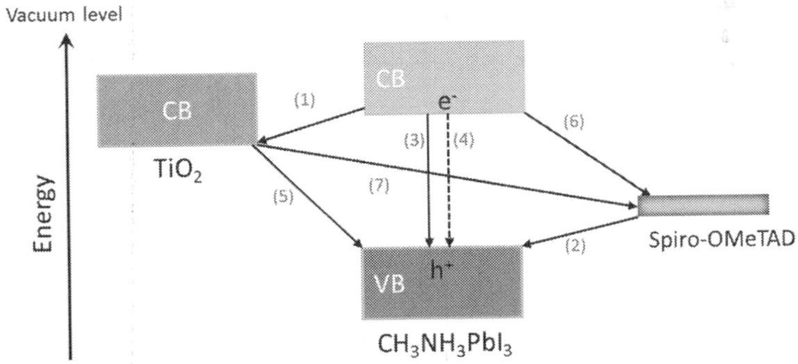

Figure 2. Schematic diagram of energy levels and electron transfer processes in an HTM/perovskite/TiO_2 cell. (1) Electron injection; (2) hole injection; (3) radiative exciton recombination; (4) non-radiative exciton recombination; (5) back electron transfer at the TiO_2 surface; (6) back charge transfer at the HTM surface; (7) charge recombination at the TiO_2/HTM interface.

With various cations/ions in the perovskite structure, there are different lattice parameters. However, it is not always suitable to have random combinations of cation A, B, and anion X.

The crystallographic stability and probable structure can be evaluated by considering the Goldschmidt's tolerance factor t and the octahedral factor μ [5], $t = (r_A + r_X)/[\text{sqrt}(2)^*(r_B + r_x)]$, where r_A, r_B, and r_X are the effective ionic radii for the ions in the A, B, and X sites, respectively. This allows us to estimate the degree of distortion of perovskite crystal structure compared to the ideal case where $t = 1$. The octahedral factor (μ) is an additional consideration for perovskite formability, where $\mu = r_B/r_X$. It has been generally accepted that the perovskite was stabilized for a tolerance factor ranging between 0.813 and 1.107 and an octahedral factor ranging from 0.442 and 0.895 [6]. It has been found that hybrid organic–inorganic devices usually degrade easily and quickly on exposure to the moisture or ultraviolet radiation [7]. Meanwhile, structural phase transformations of conventional hybrid perovskite films at different temperature ranges are also listed in **Table 1**. Still, there have been reports on the formation of single crystals when the conditions of saturation, nucleation, and growth are well controlled [8–11].

PSC	Phase	Temperature (K)	Structure	Space group	Lattice parameter (Å)
$MAPbI_3$	α	400	Tetragonal	P4*mm*	a=6.3115 b=6.3115 c=6.3161
	β	293	Tetragonal	I4*cm*	a=8.849 b=8.849 c=12.642
	γ	162–172	Orthorhombic	$Pna2_1$	a=5.673 b=5.628 c=11.182
$MAPbCl_3$	α	>178.8	Cubic	P*m*3*m*	a=5.675
	β	172.9–178.9	Tetragonal	P4/*mmm*	a=5.655 c=5.630
	γ	<172.9	Orthorhombic	$P222_1$	a=5.673 b=5.628 c=11.182
$MAPbBr_3$	α	>236.9	Cubic	P*m*3*m*	a=5.901
	β	155.1–236.9	Tetragonal	I4/*mcm*	a=8.322 c=11.833
	γ	149.5–155.1	Tetragonal	P4/*mmm*	a=5.8942 c=5.8612
	δ	<144.5	Orthorhombic	$Pna2_1$	a=7.979 b=8.580 c=11.849
$MASnI_3$	α	293.00	Tetragonal	P4*mm*	a=6.2302 b=6.2302 c=6.2316
	β	200	Tetragonal	I4*cm*	a=8.7577 b=8.7577 c=12.429
$FAPbI_3$	α	293	Trigonal	P3*m*1	a=8.9817 b=8.9817 c=11.006
	β	150	Trigonal	P3	a=17.791 b=17.791 c=10.091

Table 1. Structural phase transformations for common hybrid perovskites. The table was taken from [5] with permission.

The thickness of this sensitized layer is usually limited to be around several hundred nanometers concerning the fact that holes and electrons will be recombined quickly if the film thickness is thicker than the diffusion length of charge carriers. Therefore, we simply consider the diffusion length of excitions which depend not only on the amount but also on the concentration of precursors [12–14]. As known, the nature of ambipolar of the perovskite material leads to the transport of both electrons and holes [15]. For example, the trioxide

http://dx.doi.org/10.5772/63771

absorber $CH_3NH_3PbI_3$ has the electron–hole diffusion length of several hundred nanometers and a relatively long life time [16].

Even though hybrid perovskite films are expected to have such appealing properties, only the elaborate design of every counterpart of the whole device can make the performance of hybrid perovskite devices as great as possible. For example, the band level alignment of different layers and materials in devices should be considered carefully. As summarized from literatures, the energy bands of ETL must satisfy the following conditions [17]: (1) Its conduction band (CB) must lie under the CB of the active perovskite layer to extract electrons which reach the interfaces afterward. (2) Its valence band (VB) must lie much under the VB of the perovskite to reject the holes. (3) The electron affinity of the ETL must be greater than that of the perovskite. (4) The VB of ETL should have large difference compared to the VB of perovskite in order to reject holes, so the ETL usually has wide band gap. (5) The electron mobility in ETL must be sufficiently high. Similar requirements can also be deduced for HTL. The values of VB and CB of some commonly used materials are listed in **Figure 3**.

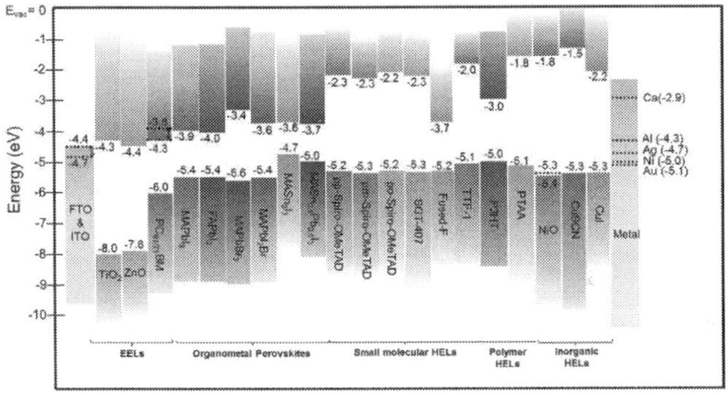

Figure 3. The VB and CB of various perovskite films. The figure was taken from [17] with permission.

The exciton property, electrical property, and the energy band of various ETL/HTL materials have been discussed so far. Moreover, optical properties will also be illustrated in this section. The mostly used technique to explore the optical property is ultraviolet-visible (UV) and photoluminescence (PL) spectroscopy. Different UV/PL spectra are shown in **Figure 4a** and **b** [18, 19]. The absorption edges of perovskite films in **Figure 4** clearly indicate that hybrid perovskite devices have high absorption factor at visible and near-infrared range, which just fit well to the solar spectrum. **Figure 5** shows the complementarities between the Si-based solar cell and the hybrid perovskite device as they have opposite external quantum efficiency (EQE). Furthermore, the internal quantum efficiency (IQE) has been reported to reach 100% through optimizing the perovskite device [20], which inspires the researchers that tandem solar cells based on perovskite, and Si is very interesting if the fabrication technology can be improved to use the sunlight more effectively.

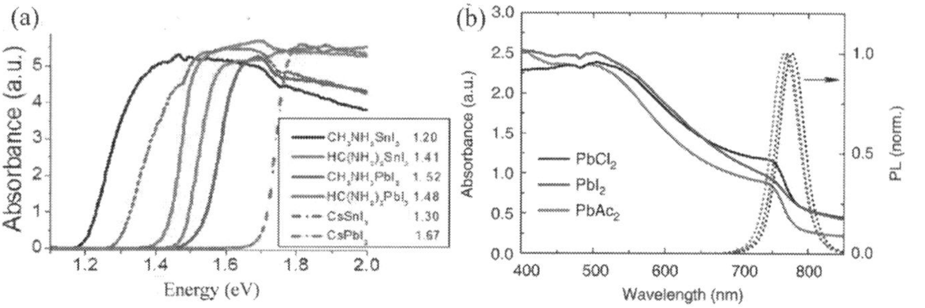

Figure 4. (a) Various sites A and B have different band gaps. The value of the band gap is given next to each compound. Figure was taken from [18] with permission; (b) UV–vis absorption spectra for perovskite films derived from different lead sources. The right axis shows the PL spectra of the perovskites prepared on glass, with photoexcitation at 507 nm. The figures are taken from [19] with permission.

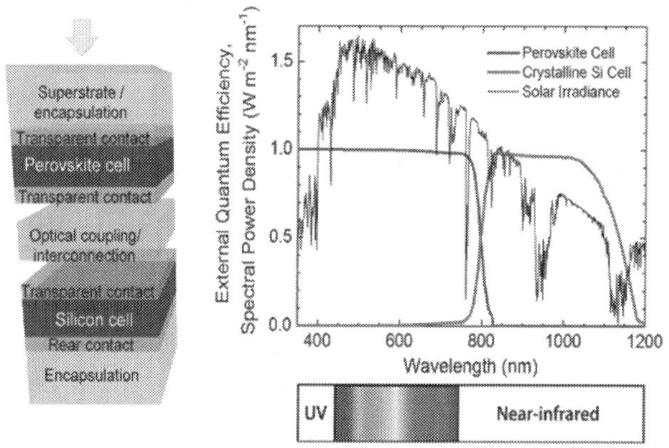

Figure 5. The complementation of EQE in Si and perovskite tandem solar cell. The image was taken from pvlab.epfl.ch/page-124775-en.html with permission.

3. Doping into different layers and the influence on devices

3.1. Doping into the electron transport layer

As discovered, the boundaries and defects existing in ETL generally result in the recombination of electrons and holes associated with pinhole/cracks between ETL and substrate and will therefore lead into the decline of PCE and stability. Consequently, proposed materials for ETL should own high electron affinity, excellent surface morphology, and effective hole-blocking properties. From this point of view, visible improvement should be realized by properly doping the ETL for the electron transport property, the hole blocking ability, since external doping usually rearranges the band alignment due to the pining of Fermi level, which may influence the electron transport property, hole blocking ability, for example, in the manner of removing deep electronic traps and sub-band states, enhance the carrier's life time and the film's conductivity by several orders of magnitude. Meanwhile, the doping into ETL can also enhance the film morphology by removing defects or filling the gap in between islands in the film. Nevertheless, the band gap of ETL itself is also modified by doping due to the change of band structures. In the following, mostly used electron transport materials are described in detail.

3.1.1. TiO_2

Rutile TiO_2 nanostructures on fluorine-doped tin oxide (FTO) substrates are interesting building blocks for solar cells (as shown in **Figure 7a**). Positive structure and FTO substrate are always adopted together for the electron transport material, since FTO can be annealed up to 450°C when rutile TiO_2 is formed. The properties of large band gap, suitable band edge for charge injection and extraction, long lifetime of excited charge carriers, exceptional resistance to photo corrosion, non-toxicity, and low cost have made TiO_2 a popular material for solar energy applications. In spite of this, pure TiO_2 is not ideal for ETL which need to be doped to achieve n-type characters for higher electron extraction.

The scale of the pure TiO_2 nanopaticles which results in the highest PCE in organic–inorganic devices is about 50 nm [21]. The crafts for the fabrication of compact TiO_2 layer (spin coating, spray pyrolysis, sol-gel methods, magnetron sputtering) have developed rapidly, and one effective way is to decorate TiO_2 in order to improve the device's performance. For example, Mg-doped TiO_2 can increase the open-circuit voltages (V_{OC}) originated from elevating CB by a microwave hydrothermal reaction [22]. Yttrium-doped TiO_2 (Y-TiO_2) was used as the ETL to enhance electron extraction and transport [23]. Zr/Pyridine-doped TiO_2 could reduce the hysteresis and improve the performance of organic perovskite device by increasing the carrier's lifetime [24]. Nb-doped TiO_2 reduced selectively the contact resistance and increased the charge recombination resistance, as revealed by the impedance spectroscopy measurements [25]. A low level of Al doping into TiO_2, by adding the aluminum isopropoxide precursor into Ti isopropoxide solution, can reduce non-stoichiometric oxygen-induced defects in TiO_2 layer. As discovered, the substitution/doping removes deep electronic traps and sub-band states, enhances the conductivity by several orders of magnitude, and improves the stability of corresponding devices. In all, metal doping of TiO_2 seems to be the main method, while other approaches of doping also exist, for example, TiO_2 mixed with graphene, which promises the low cost as the compact ETL synthesized by solution-based deposition procedure [26]. Another doping-like method is inserting ultrathin graphene quantum dots (GQDs) between TiO_2 and perovskite layer, which could reduce the electron extraction time from 300 to 100 ps [27]. Nevertheless, we should bear in mind that most of these doping methods have been utilized in dye-sensitized solar cells. Traps appear after doping in TiO_2, and new energy levels are formed. As can be seen in **Figure 6**, the doping concentration determines the operation mechanism of electron and affects the V_{OC} and short-circuit current (J_{SC}) [28].

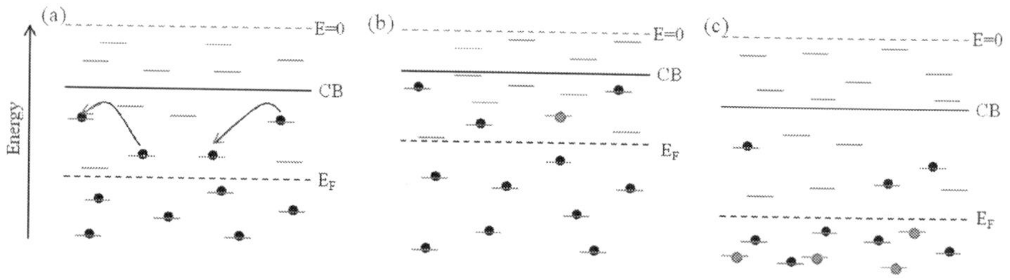

Figure 6. Illustration of the doping effect on the CB and E_F in TiO_2. E_F is the Fermi level, and the doping induced states are also displayed in green. (a) In pristine TiO_2, electrons are transported by "hopping" from shallow trap to shallow trap until getting to the electrode. Therefore, the electron transport rate and J_{SC} will be affected by the density of shallow trap. Deep traps can permanently trap electrons and act as recombination sites, affecting V_{OC}. Another important factor determining V_{OC} is E_F, as V_{OC} is defined as the difference between E_F of TiO_2 and the HTM. (b) The elimination of the deep trap density, resulting in an upward shift of E_F and thus an increase of V_{OC}. Because the CB is shifted toward the CB of the absorber, the driving force for electron injection is lowered, in combination with a decreased trap density and the related electron transport, lowering J_{SC}. (c) In the case of formation of deep traps by doping, the CB and E_F are shifted downward. Together with the enhanced recombination through the deep trap states V_{OC} decreases. Due to the larger offset between the CB and the absorber CB, electron injection is improved and the higher trap density causes an increase in electron transport, resulting in an enhanced J_{SC}. Ideally, the dopant eliminates deep traps while introducing new states close to CB, enhancing both V_{OC} and J_{SC} through decreased recombination and increased electron transport.

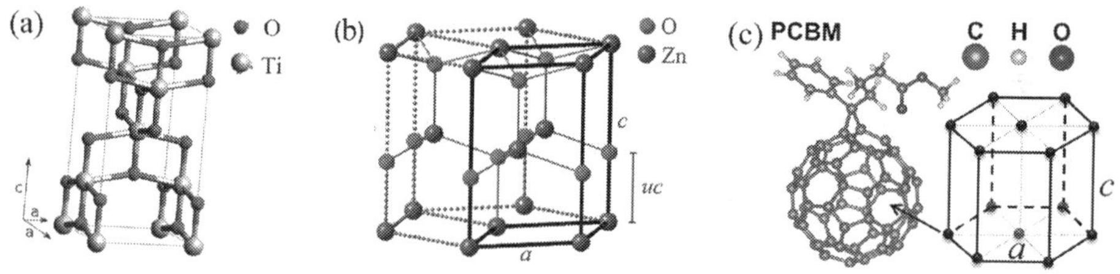

Figure 7. (a) The unit cell of TiO_2; oxygen atoms form a distorted octahedron with a titanium atom at the center; (b) the ZnO unit cell; (c) the molecular drawing of the PCBM with the unit cell labeled.

On the contrary, the size of the particles used as mesoporous TiO_2 is very small (around 15 nm) and can be easily controlled by tuning the concentration, pH of the reactants, and the annealing temperature. As also discovered, reduced graphene oxide/mesoporous TiO_2 nanocomposite as scaffold material could improve the electron collection efficiency, enhance the V_{OC} and J_{SC}, and reduce the interface resistance [29]. Doping into the mesoporous TiO_2 has been tried by self-assembling one monolayer of fullerene (C_{60}) to get a functionalized mesoporous titania for good electron extraction [30].

3.1.2. ZnO

Compared to TiO_2, ZnO (**Figure 7b**) owing to high electron collection as ETL is more environment friendly as it does not need to be sintered as high as to 450°C to become rutile crystal and grows very fast by solution process [31]. Furthermore, less chance of recombination of carriers would occur in ZnO because its conductivity is several orders of magnitude higher than that of TiO_2 [32–34]. As known, ZnO is an important II–VI semiconductor with a direct band gap of 3.4 eV and an exciton binding energy of 60 meV at room temperature. Moreover, ZnO has excellent optical properties such as a high infrared reflectivity and transparency in the visible spectrum. In addition, it is abundant in nature and inexpensive [35]. However, ZnO accelerates the degradation of perovskite layer into PbI_2 when it is thermally annealed [36]. Similar to TiO_2, ZnO also suffers from photo corrosion which reduces the performance and stability [37].

As reported, Al-doped ZnO lifted the PCE from 10.8 to 12.0% by optimizing the thickness of ETL and smoothing the interface [38]. Though fewer reports have been found for the doping of ZnO till date, it is proved gradually to be an excellent candidate with external doping to replace TiO_2 and to solve the problems existing in TiO_2. In a word, metal atoms are always chosen to be the doping materials in the doping of either TiO_2 or ZnO.

3.1.3. PCBM

PCBM (**Figure 7c**) is one of the fullerene derivatives. Its excellent properties of high electron affinity and transmission make it extremely popular in organic solar cells. PCBM is widely used in inverted perovskite structure accompanied with indium tin oxide (ITO) substrate.

PCBM doped with 1,3-dimethyl-2-phenyl-2,3-dihydro-1H-benzoimidazole (DMBI) results in n-doped ETL which significantly increases the J_{SC} [39], while oleamide-doped PCBM improves the property of electron transport [40]. It has also been reported that PCBM doped with graphdiyne [41], and the inverted hybrid perovskite structure spun with PCBM/C_{60} could reduce the traps effectively which passivates the charge trap states and eliminates the notorious photocurrent hysteresis [42, 43]. Moreover, PCBM doped with PFNOX and polystyrene (PS) was found to enhance J_{SC}, while negligible hysteresis was discovered due to PFNOX which reduced the electron recombination [44]. The advantages of PCBM compared with metal-oxide ETL are that it can be used in soft hybrid perovskite devices with soluble process at the cost of low power consumption.

3.2. Doping into the perovskite layer

3.2.1. Doping at site A

The cation at site A has no direct contribution to the properties of electron injection or transport. The B–X octahedra and A are relatively independent owing to the fact that there is almost no overlap between the electron clouds of A and B-X due to the large unit cell [45]. However, the radius of ion A affects the symmetry of the perovskite structure which in turn affects the electron's properties [46–48]. As summarized from literatures, site A is always occupied by Cs, Rb, methylammonium (MA), formamidinium (FA), or ethylammonium (EA). The first study about mixed cations $(MA)_x(FA)_{1-x}PbI_3$ claimed that the PCE reaches up to 14.9% due to the changing band gap which facilities the photon absorption much more effectively [49]. Afterward, $(MA)_x(FA)_{1-x}PbI_3$-based hybrid devices have been fabricated with rather high PCE [46,

48, 50]. **Figure 8a** and **b** shows the difference in emission spectrum and the change of light harvesting due to different ratios of FAI and MAI in hybrid perovskite solar cells. Similar discoveries were also reported by our group when doping MA with small amount of CH_3SH organics, while the PCE increased from 6.9 to 9.4% [51], as presented in **Figure 8c**. When site A is doped by Cs with optimized doping concentration and film thickness, the PCE reaches over 10%, and the film morphology was found to be more compact and uniform [52]. For example, PCE over 17% was reported for the $Cs_{0.2}FA_{0.8}PbI_{2.84}Br_{0.16}$ hybrid device as well as the excellent stability in ambient air compared to $FAPbI_3$, which was attributed to uniform coating on TiO_2 and no separation in chemical phases [53]. Methylammonium(MA) iodide and 5-ammoniumvaleric acid (5-AVA) iodide hybrid perovskite device was also found to have high stability of perovskite structure and performance [54], which can be attributed to the lower defect concentration and better pore filling as well as complete contact with the mesoporous scaffold, as illustrated in **Figure 8d**.

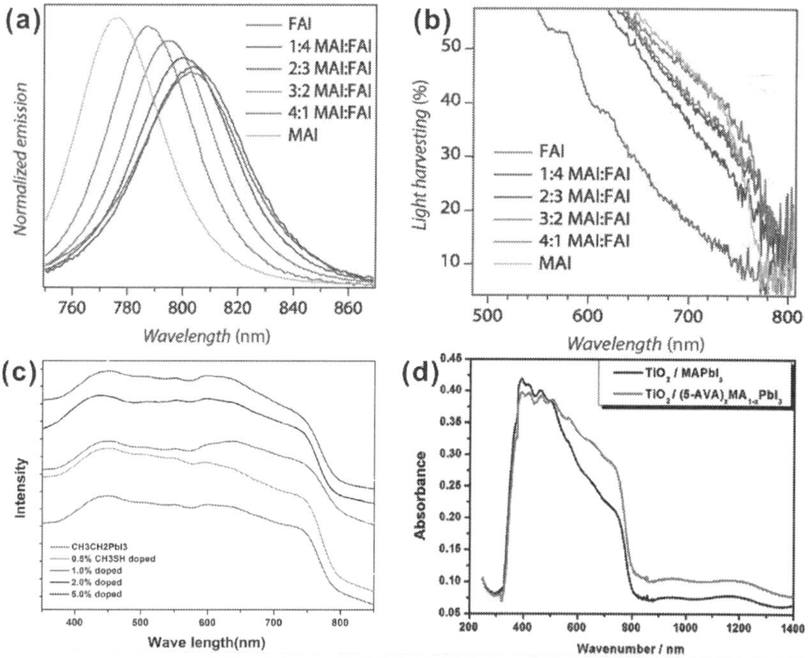

Figure 8. (a) Normalized emission of $(CH_3NH_3)_x(HNCHNH_3)_{1-x}PbI_3$ (x = 0, 0.2, 0.4, 0.6, 0.8, 1). The emission is shifted and broadened as a function of x. (b) Light-harvesting spectra of $(CH_3NH_3)_x(HNCHNH_3)_{1-x}PbI_3$ films with different x. As the concentration of formamidinium was increased, the absorbance of the film decreased, while no change in the band gap was discernable. (c) UV absorption spectra of $(CH_3SH_2)_x(CH_3NH_3PbI_3)_{1-x}$ where the changes of band gap are visible. (d) The UV–vis spectra of FTO glass/TiO_2 films filled with $(5\text{-}AVA)_xMA_{1-x}PbI_3$ and $MAPbI_3$. Some images were taken from [49] and [54] with permission.

3.2.2. *Doping at site B*

The ideal element which suits the organic–inorganic hybrid perovskite structure at site B is Pb, and one can easily refer to the same group elements, for example, Ge or Sn. Javier Navas and his coworkers tried to dope/replace Pb^{2+} with Sn^{2+}, Sr^{2+}, Cd^{2+}, and Ca^{2+} and revealed the development of the properties of crystalline phase, band structure (**Figure 9a**), and emission and optical characteristics. Non-covalent interaction (NCI) analysis complemented with the results from electron localization functions (ELFs) indicated that the strength of the dopant-I interaction follows the order with Cd-I > Sn-I ≈ Pb-I > Sr-I for the tetragonal structure [55]. The system of $MASn_{1-x}Pb_xI_3$ was explored in detail at different ratios of Sn and Pb (**Figure 9b**) [18, 56], and optimized hybrid devices were fabricated afterward [57, 58]. In the end, it is worth pointing out that Pb is rather toxic, and environmental friendly materials are urgently needed to replace lead at site B. As far as explored, no other stable element has been found which can fully replace Pb without downgrading the hybrid perovskite devices' performance. Consequently, doping at site B seems to be the dominant approach to resolve the above-mentioned challenges.

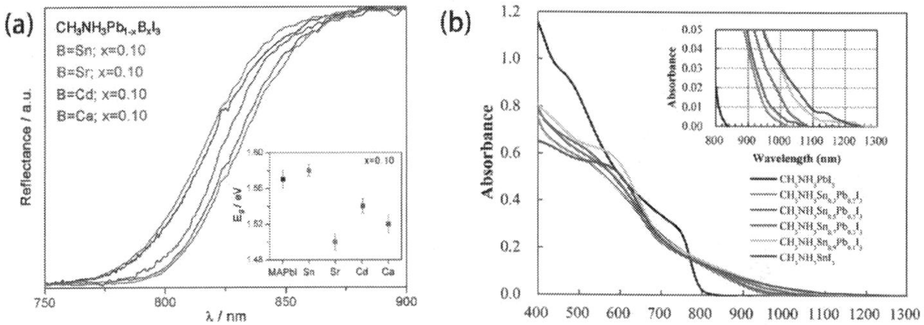

Figure 9. (a) Diffuse reflectance UV–vis spectra for the $MAPbI_3$ sample and $MAPb_{1-x}B_xI_3$, with B = Sn^{2+}, Sr^{2+}, Cd^{2+}, Ca^{2+}, and x = 0.10. In the inset, the plot of the optical band gap values for each sample. (b) Electronic absorption spectra of perovskite $CH_3NH_3Sn_xPb_{(1-x)}I_3$ coated on porous TiO_2. Images were taken from [55] and [56] with permission.

3.2.3. *Doping at site X*

Site X, usually occupied by halogen atoms, typically was iodine at beginning and was attempted with doping of Cl afterward which encouragingly claimed the improvement of both diffusion length and carrier lifetime [14, 58]. Doped $MAPbI_{3-x}Cl_x$ has highly oriented crystalline structures exemplified through the strong (1 1 0), (2 2 0), and (3 1 0) peaks attributed to the tetragonal phase [59, 60]. As discovered previously, the variation of the film crystalline orientation order does not impose effect on the photovoltaic performance of the hybrid perovskite devices. [61]. The influence of doping of Cl is not only on the crystalline structure but also on its formation route of perovskite structure. As reported, doping a little amount of Cl in $MAPbI_3$ perovskite film serves as the nucleation sites to form suitable surface coverage of perovskite film [62]. Moreover, $MAPbI_{3-x}Cl_x$-based hybrid solar cell changes both the phase and optical properties by optimizing the annealing temperature [63], as also discovered by our group recently with an *in situ* investigation of the improvement of structure and performance of $MAPbI_{3-x}Cl_x$perovskite device by annealing [64], as present in **Figure 10a** and **b**. In most instances, optimized annealing temperature will enhance the PCE in the Cl-doped system, although the PCE may also be reduced after proper annealing in a few cases [65], since the performance of perovskite solar cell depends strongly on the preparation of film formation and crystallization [66, 67]. Interestingly, planar structure comprising $CH_3NH_3PbI_{3-x}Cl_x$ film exhibits the p-doping character and therefore, a p-n heterojunction was formed when contacted with the n-doped TiO_2 compact layers [68]. The formation energy of mixed halide perovskite was determined by annealing temperature as also verified by theoretical calculations [69].

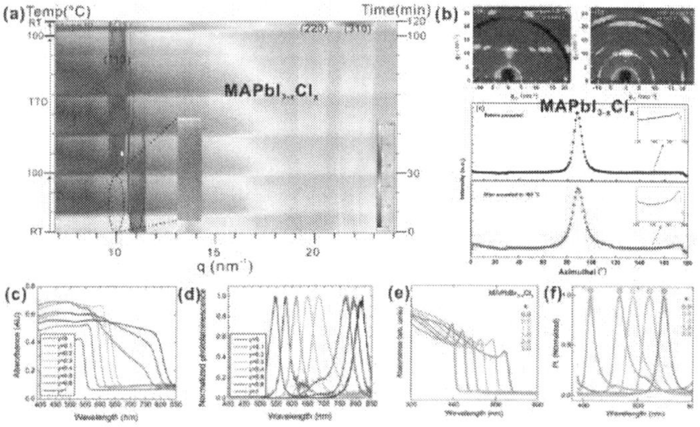

Figure 10. (a) The tunability of band gap and structure in the mixed halide perovskite system.(b) The perovskite structure transformation for the Cl-doped MAPbI3 film during programmed annealing investigated with conventional and two dimensional XRD. (c) UV–vis of the $FAPbI_yBr_{3-y}$ perovskites with varying y. (d) The related photoluminescence spectra for the same films. (e) and (f) Optical absorbance and photoluminescence spectra (370 nm excitation) of $MAPbBr_{3-x}Cl_x$ films. Figure (c, d, e, f) were taken from [47] and [73] with permission.

Apart from the doping of Cl, Br has also been tried to partly replace I at the site X. In fact, bromide has been effectively used to tune the band gap of perovskite films. UV and PL spectrum are shown for various doping of Br from 0 to 100% in $CH_3NH_3PbI_3$ perovskite films in **Figure 10c** and **d** [47]. For the synthesis the perovskite $CH_3NH_3PbI_{3-x}Br_x$, doping the MAI and PbI_2 solution with MABr can change the band gap, which overlaps the whole usable solar spectrum for perovskite solar cells and shifts the XRD peak due to the different ratio of MABr and MAI [70]. The calculation for mixed halide has also been done using first principle calculations. For example, Herz et al. studied the charge carrier dynamics and mobility in formamidinium lead-mixed-halide perovskites and found that the auger recombination constant exists in the mixed halide perovskite system which strongly depends on the value of y in $FAPb(Br_yI_{1-y})_3$ [71]. Besides, lowering the energetic disorder in mixed-halide perovskites would significantly improve charge carrier transport, allowing effective incorporation in planar-heterojunction tandem solar cells with high short-circuit currents, open-circuit voltages, and PCEs [71]. Perovskite solar cell based on $CH_3NH_3SnI_{3-x}Br_x$ has been fabricated as a lead-free hybrid device with PCE getting to 5.73% [72]. The last combination of doping with halides can be Cl and Br. $CH_3NH_3Pb(Br_xCl_{1-x})_3$ was used as different types of light emitting diodes (LEDs) due to different energy gaps induced from different doping ($0 < x < 1$) [73, 74]. Such $CH_3NH_3Pb(Br_{1-x}Cl_x)_3$-based hybrid devices with different inclusion of Cl exhibit different electronic structures and the band gap gets broadened after the inclusion of Cl [75]. The optical properties are shown, respectively, in **Figure 10e** and **f**.

Another choice of doping at the site X could be the utilization of anion. $MAPbI_{3-x}(BF_4)_x$ has a band gap of 1.5 eV slightly lower than that of $MAPbI_3$ and the absorption wavelength is around 827 nm, where the sunlight spectrum has stronger irradiance compared to the absorption wavelength of 800 nm for $MAPbI_3$. $MAPbI_{3-x}(BF_4)_x$ was proved to own good properties such as good electrical conductivity and high photo response [76]. $MAPb(SCN)_2I$ was demonstrated to show high stability in moisture environment compared to $MAPbI_3$, and the fabrication process is rather similar [77].

3.2.4. Mixed doping

The $MAPbI_3$–PEOXA precursor was used as the sensitizer of hybrid perovskite solar cell to improve the interface morphology of the perovskite-polymer films and to reduce the undesired contact between PEDOT:PSS and PCBM layers for the minimized shunting path at the device [78]. It can be concluded now that the main point for doping hybrid perovskite device is that the proper parameters of t, μ, nice morphology, and robust structure to moisture as well as high performance should be ensured when faced with various choices of doping.

3.3. HTL doping

3.3.1. NiOx

NiO (as shown in **Figure 11a**) is a relatively rare material which has a nature of p-type semiconductor in both perovskite device and dye-sensitized solar cells, so it is a promising candidate for hybrid devices. The NiO_x thin film is very compact without pinholes which can block the electrons very effectively [79]. Therefore, it is expected that NiO_x doping can improve the properties of hybrid organic inorganic devices. For example, Cu-doped NiO_x increases the electron transmission significantly and enhances the PCE from 8.9 to 15.4% in the inverted structured solar cell [80]. Afterward, low-temperature process for the doping of Cu has been reported [81]. $Ni_xMg_{1-x}O$ was also used as HTL and led to large-area and stable devices with higher PCE [82]. Therefore, we can say that the NiOx will replace the organic HTL due to its commercial future.

Moreover, graphene oxide (GO) was proved to be a promising candidate as HTL, owing to the enhanced crystallization, high surface coverage ratio, and preferred in-plane orientation of the (110) plane of perovskite film on top. Efficient hole extraction and the enhanced stability of hybrid devices were demonstrated in the GO-based inverted devices, as reported by our group [83] and other studies [84]. CuI is another inorganic material used as HTL which is a p-type semiconductor and has a matched band structure with the perovskite films [85].

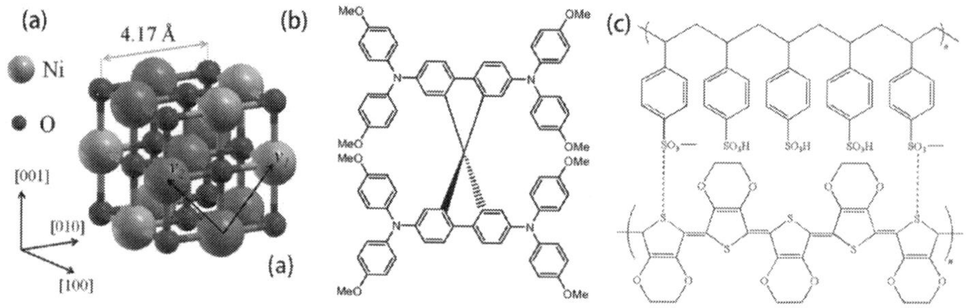

Figure 11. (a) Structure of NiO; (b) structure of spiro-OMeTAD; (c) structure of PEDOT:PPS.

3.3.2. Spiro-MeOTAD and PEDOT:PPS

Spiro-MeOTAD (shown in **Figure 11b**) was also used as HTL as an alternate which supports great PCE for the fabricated devices [86, 87]. However, all of the Spiro-MeOTAD based layers need to be doped because of the nature of such material with bad charge transmission. For instance, Spiro-OMeTAD doped with iodide-reduced graphene oxide has increased the stability of corresponding device and reduced the cost of fabrication. Spiro-MeOTAD116 was doped with Co(III) complex as a p-type dopant for the HTL to ensure a sufficient conductivity and low series resistance [88, 89]. The n-type doped Spiro-OMeTAD functionalized as the pinhole-free HTL has a proper band structure which matches well the band of perovskite film [90].

Moreover, dopant-free PBDTTT-C polymer [91], dopant-free spiro-CPDT [92], dopant-free TPB [93], and dopant-free linear acene derivative were also utilized as the hole transport material [94]. These materials are used as HTL for their appealing properties to replace Spiro-OMeTAD.

PEDOT: PPS (shown in **Figure 11c**) is always designed as the HTL in inverted hybrid perovskite devices, but its acidic characteristics severely threaten long-term stability and performance [95]. Thus, the substitute was explored and found and, for example, CPE-K which contributes to a better performance of the related device [96], is gradually used as HTL. In fact, graphene can not only be used in ETL but also in HTL. PEDOT: PPS doped with graphene quantum rings (GQRs) ensures the efficient hole extraction in hybrid solar cells and relatively high PCE. Hydrophilic graphene oxide doping in PEDOT: PSS composited films demonstrated a PCE of 1.8 times higher than that of the hybrid electronics based on the pristine PEDOT: PSS buffer layer [97]. Besides, PTAA (Poly[bis(4-phenyl)(2,5,6-trimentlyphenyl)amine, P3HT(poly(3-hexylthiophene)), PCPDTBT (Poly[2,6-(4,4-bis-(2-ethylhexyl)-4H-cyclopenta [2,1-b;3,4-b′]dithiophene)-alt-4,7(2,1,3-benzothiadiazole)]), VB-DAAF (vinylbenzyl 9,9-

http://dx.doi.org/10.5772/63771

diarylfluorene-based triaryldiamine), etc. are also used as HTL for their cheap cost, easy synthesis, and better performance. Actually, the fabricated devices showed higher PCE and better performance than the commonly used PEDOT:PPS and Spiro-MeOTAD materials [98–101].

4. Conclusion

In conclusion, the physical structure and electronic and optical properties of hybrid organic–inorganic perovskite electronics have been illustrated in detail, which sheds new light on the implication of solar cell industry. Driven by the urgent concerns from hybrid perovskite device, doping into the perovskite solar cell is explored by researchers in order to get stable devices with high performance. The progress of *in situ* doping of perovskite electronics from worldwide has been reviewed in this chapter, including the achievement from our group. The doping of ETL promotes the electron transmission, enhances the surface smoothness, blocks holes effectively, reduces the traps, and promotes the PCE for the device ultimately. Second, the doping of perovskite film itself is discussed regarding the doping at sites A, B, and X. Doping at site A leads to a higher PCE, which is associated with better absorption of sunlight according to the UV spectrum. Doping at site B mainly contributes to the environment friendliness with the final goal to get lead-free hybrid devices. Doping at site X can easily tune the band gap of perovskite films to absorb more sunlight with different wavelengths, making the interface more suitable for the electron transmission. Organic HTL (Spiro-MeOTAD) is expensive and needs to be doped to become a p-type semiconductor, so new kind of Spiro-MeOTAD-like materials are synthesized at low cost without doping to improve the hole blocking and electron transmission. Another organic HTL–PEDOT:PPS is also discussed due to its wide usage and easy fabrication by solution process. In all, the PCE of the hybrid perovskite solar cell increases rapidly one study by one study, while the doping promotes the performance of such hybrid organic–inorganic electronics, which is expected to be the third generation of solar cell in the near future.

Author details

Kongchao Shen[1,2*], Hao Liang Sun[1], Gengwu Ji[1], Yingguo Yang[1], Zheng Jiang[1] and Fei Song[1]

*Address all correspondence to: songfei@sinap.ac.cn

1 Shanghai Institute of Applied Physics, Chinese Academy of Sciences, Shanghai, P. R. China

2 Department of Physics, Zhejiang University, Hangzhou, P. R. China

References

[1] Kojima, A., et al. Organometal halide perovskites as visible-light sensitizers for photovoltaic cells. J Am Chem Soc. 2009; 131:6050.

[2] Bi, D., et al. Efficient luminescent solar cells based on tailored mixed-cation perovskite. Sci Adv. 2016; 2: e1501170.

[3] Sum, T. C., et al. Advancements in perovskite solar cells: photophysics behind the photovoltaics. Energy Environ Sci. 2014; 7: 2518.

[4] Ponseca, C. S., et al. Organometal halide perovskite solar cell materials rationalized: ultrafast charge generation, high and microsecond-long balanced mobilities, and slow recombination. J Am Chem Soc. 2014; 136: 5189.

[5] Li, C., et al. Formability of ABX3 (X= F, Cl, Br, I) halide perovskites. Acta Crystallogr B. 2008; 64: 702.

[6] Kim, H.-S., et al. Organolead halide perovskite: new horizons in solar cell research. J Phys Chem C. 2014; 118: 5615–5625.

[7] Green, M. A., et al. The emergence of perovskite solar cells. Nat Photon. 2014; 8: 506.

[8] Chen, Q., et al., Under the spotlight: the organic–inorganic hybrid halide perovskite for optoelectronic applications. Nano Today. 2015; 10: 355.

[9] Liu, Y., et al., Two-inch-sized perovskite $CH_3NH_3PbX_3$(X = Cl, Br, I) crystals: growth and characterization. Adv Mater. 2015; 27: 5176.

[10] Huang, J., et al. Organometal trihalide perovskite single crystals: a next wave of materials for 25% efficiency photovoltaics and applications beyond? J Phys Chem Lett. 2015; 6: 3218.

[11] Saidaminov, M. I., et al. High-quality bulk hybrid perovskite single crystals within minutes by inverse temperature crystallization. Nat Commun. 2015; 6: 7586.

[12] Eperon, G. E., et al. Morphological control for high performance, solution-processed planar heterojunction perovskite solar cells. Adv Funct Mater. 2014; 24: 151.

[13] Etgar, L., et al., Mesoscopic $CH_3NH_3PbI_3/TiO_2$ heterojunction solar cells. J Am Chem Soc. 2012; 134: 17396.

[14] Stranks, S. D., et al., Electron–hole diffusion lengths exceeding 1 micrometer in an organometal trihalide perovskite absorber. Science. 2013; 342: 341.

[15] Ball, J. M., et al. Low-temperature processed meso-superstructured to thin-film perovskite solar cells. Energy Environ Sci. 2013; 6: 1739.

[16] Dong, Q. F., et al. Electron–hole diffusion lengths > 175 μm in solution-grown $CH_3NH_3PbI_3$ single crystals. Science. 2015; 347: 967–970.

[17] Chueh, C., et al. Recent progress and perspective in solution-processed interfacial materials for efficient and stable polymer and organometal perovskite solar cells. Energy Environ Sci. 2015; 8: 1160.

[18] Stoumpos, C. C., et al. Semiconducting tin and lead iodide perovskites with organic cations: phase transitions, high mobilities, and near-infrared photoluminescent properties. Inorg Chem. 2013; 52: 9019.

[19] Zhang, W., et al. Ultrasmooth organic–inorganic perovskite thin-film formation and crystallization for efficient planar heterojunction solar cells. Nat Commun. 2015; 6: 6142.

[20] Yang, B., et al. Perovskite solar cells with near 100% internal quantum efficiency based on large single crystalline grains and vertical bulk heterojunctions. J Am Chem Soc. 2015; 137: 9210.

[21] Sung, S. D., et al. 50 nm sized spherical TiO_2 nanocrystals for highly efficient mesoscopic perovskite solar cells. Nanoscale. 2015; 7: 8898.

[22] Manseki, K., et al. Mg-doped TiO_2 nanorods improving open-circuit voltages of ammonium lead halide perovskite solar cells. RSC Adv. 2014; 4: 9652.

[23] Zhou, H., et al. Interface engineering of highly efficient perovskite solar cells. Science. 2014; 345: 542.

[24] Nagaoka, H., et al. Zr Incorporation into TiO_2 electrodes reduces hysteresis and improves performance in hybrid perovskite solar cells while increasing carrier lifetimes. J Phys Chem Lett. 2015; 6: 669.

[25] Yin, X., et al. Performance enhancement of perovskite-sensitized mesoscopic solar cells using Nb-doped TiO_2 compact layer. Nano Res. 2015; 8: 1997.

[26] Wang, J. T., et al. Low temperature processed electron collection layers of graphene / TiO_2 nanocomposites in thin film perovskite solar cells. Nano Lett. 2014; 14: 724.

[27] Zhu, Z., et al. Efficiency enhancement of perovskite solar cells through fast electron extraction: the role of graphene quantum dots. J Am Chem Soc. 2014; 136: 3760.

[28] Roose, B., et al. Doping of TiO_2 for sensitized solar cells. Chem Soc Rev. 2015; 44: 8326.

[29] Han, G. S., et al. Reduced graphene oxide/mesoporous TiO_2-nanocomposite based perovskite solar cells. ACS Appl Mater Interface. 2015; 7: 23521.

[30] Abrusci, A., et al. High-performance perovskite-polymer hybrid solar cells via electronic coupling with fullerene monolayers. Nano Lett. 2013; 13: 3124.

[31] Zhang, J., et al. Fast and low temperature growth of electron transport layers for efficient perovskite solar cells. J Mater Chem A. 2015; 3: 4909.

[32] Mahmood, K., et al. Double-layered ZnO nanostructures for efficient perovskite solar cells. Nanoscale. 2014; 6: 14674.

[33] Magne, C., et al. Effects of ZnO film growth route and nanostructure on electron transport and recombination in dye-sensitized solar cells. J Mater Chem A. 2013; 1: 2079.

[34] Law, M., et al. Nanowire dye-sensitized solar cells. Nat Mater. 2005; 4: 455.

[35] Mahmood, K., et al. Double-layered ZnO nanostructures for efficient perovskite solar cells. Nanoscale. 2014; 6: 14674.

[36] Cheng, Y., et al. Decomposition of organometal halide perovskite films on zinc oxide nanoparticles. ACS Appl Mater Interface. 2015; 7: 19986.

[37] Zhang, H., et al. Photocorrosion inhibition and photoactivity enhancement for zinc oxide via hybridization with monolayer polyaniline. J Phys Chem C. 2009; 113: 4605.

[38] Peng, G., Xu, X. & Xu, G. Hybrid organic–inorganic perovskites open a new era for low-cost, high efficiency solar cells. J Nanomater. 2015; 1: 241853.

[39] Kim, S. S., et al. Performance enhancement of planar heterojunction perovskite solar cells by n-doping of the electron transporting layer. Chem Commun. 2015; 51: 17413.

[40] Xia, F., et al., Efficiency enhancement of inverted structure perovskite solar cells via oleamide doping of PCBM electron transport layer. ACS Appl Mater Interface. 2015; 7: 13659.

[41] Kuang, C., et al., Highly efficient electron transport obtained by doping PCBM with graphdiyne in planar-heterojunction perovskite solar cells. Nano Lett. 2015; 15: 2756.

[42] Shao, Y., Xiao, Z., Bi, C., Yuan, Y. & Huang, J. Origin and elimination of photocurrent hysteresis by fullerene passivation in $CH_3NH_3PbI_3$ planar heterojunction solar cells. Nat Commun. 2014; 5: 5784.

[43] Xu, J., et al. Perovskite–fullerene hybrid materials suppress hysteresis in planar diodes. Nat Commun. 2015; 6: 7081.

[44] Zhu, Z., et al. A PCBM electron transport layer containing small amounts of dual polymer additives that enables enhanced perovskite solar cell performance. Adv Sci. 2015; DOI: 10.1002/advs.201500353.

[45] Borriello, I., Cantele, G. & Ninno, D., Ab initio investigation of hybrid organic–inorganic perovskites based on tin halides. Phys Rev B. 2008; 77: 235214.

[46] Pang, S., et al. $NH_2CH{=}NH_2PbI_3$: an alternative organolead iodide perovskite sensitizer for mesoscopic solar cells. Chem Mater. 2014; 26: 1485.

[47] Eperon, G. E., et al. Formamidinium lead trihalide: a broadly tunable perovskite for efficient planar heterojunction solar cells. Energy Environ Sci. 2014; 7: 982.

[48] Aharon, S., et al. Temperature dependence of hole conductor free formamidinium lead iodide perovskite based solar cells. J Mater Chem A. 2015; 3: 9171.

[49] Pellet, N., et al. Mixed-organic-cation perovskite photovoltaics for enhanced solar-light harvesting. Ang Chem Inter Ed. 2014; 53: 3151.

[50] Lee, J., Seol, D., Cho, A. & Park, N., High-efficiency perovskite solar cells based on the black polymorph of $HC(NH_2)_2PbI_3$. Adv Mater. 2014; 26: 4991.

[51] Li, H., et al. Performance improvement of $CH_3NH_3PbI_3$ perovskite solar cell by CH_3SH doping. Nanomater Nanotechnol. 2016; 6: 24.

[52] Choi, H., et al. Cesium-doped methylammonium lead iodide perovskite light absorber for hybrid solar cells. Nano Energy. 2014; 7: 80.

[53] Yi, C., et al. Entropic stabilization of mixed A-cation ABX_3 metal halide perovskites for high performance perovskite solar cells. Energy Environ Sci. 2016; 9: 656–662.

[54] Mei, A., et al. A hole-conductor-free, fully printable mesoscopic perovskite solar cell with high stability. Science. 2014; 345: 295.

[55] Navas, J., et al. New insights into organic–inorganic hybrid perovskite $CH_3NH_3PbI_3$ nanoparticles. An experimental and theoretical study of doping in Pb^{2+} sites with Sn2+, Sr^{2+}, Cd^{2+} and Ca^{2+}. Nanoscale. 2015; 7: 6216.

[56] Ogomi, Y., et al. $CH_3NH_3Sn_xPb_{(1-x)}I_3$ Perovskite solar cells covering up to 1060 nm. J Phys Chem Lett. 2014; 5: 1004.

[57] Noel, N. K., et al. Lead-free organic–inorganic tin halide perovskites for photovoltaic applications. Energy Environ Sci. 2014; 7: 3061.

[58] Hao, F., Stoumpos, C. C., Cao, D. H., Chang, R. P. H. & Kanatzidis, M. G. Lead-free solid-state organic–inorganic halide perovskite solar cells. Nat Photonics. 2014; 8: 489.

[59] Niu, G., et al. Review of recent progress in chemical stability of perovskite solar cells. J Mater Chem A. 2015; 3: 8970.

[60] Unger, E. L., et al. Chloride in lead chloride-derived organo-metal halides for perovskite-absorber solar cells. Chem Mater. 2014; 26: 7158.

[61] Guerra, V. L. P., et al. Implications of TiO_2 surface functionalization on polycrystalline mixed halide perovskite films and photovoltaic devices. J Mater Chem A. 2015; 3: 20811.

[62] Tidhar, Y., et al. Crystallization of methyl ammonium lead halide perovskites: implications for photovoltaic applications. J Am Chem Soc. 2014; 136: 13249.

[63] Luo, D., et al. Cubic structure of the mixed halide perovskite $CH_3NH_3PbI_{3-x}Cl_x$ via thermal annealing. RSC Adv. 2015; 5: 85480.

[64] Yang, Y. G, et al. Annealing induced structural transition and photovoltaic performance of perovskite: an in-situ real-time investigation. (submitted to ACS Nano)

[65] Khatiwada, D., et al. Efficient perovskite solar cells by temperature control in single and mixed halide precursor solutions and films. J Phys Chem C. 2015; 119: 25747.

[66] Docampo, P., et al. Solution deposition-conversion for planar heterojunction mixed halide perovskite solar cells. Adv Energy Mater. 2014; 4: 1.

[67] Barrows, A. T., et al. Efficient planar heterojunction mixed-halide perovskite solar cells deposited via spray-deposition. Energy Environ Sci. 2014; 7: 2944.

[68] Guerrero, A., et al. Electrical field profile and doping in planar lead halide perovskite solar cells. Appl Phys Lett. 2014; 105: 133902.

[69] Yin, W., et al. Anomalous alloy properties in mixed halide perovskites. J Phys Chem Lett. 2014; 5: 3625.

[70] Noh, J. H., et al. Chemical management for colorful, efficient, and stable inorganic–organic hybrid nanostructured solar cells. Nano Lett. 2013; 13: 1764.

[71] Rehman, W., et al. Charge-carrier dynamics and mobilities in formamidinium lead mixed-halide perovskites. Adv Mater. 2015; 27: 7938.

[72] Hao, F., et al. Lead-free solid-state organic–inorganic halide perovskite solar cells. Nat Photonics. 2014; 8: 489.

[73] Comin, R., et al. Structural, optical, and electronic studies of wide-bandgap lead halide perovskites. J Mater Chem C. 2015; 3: 8839.

[74] Sadhanala, A., et al. Blue-green color tunable solution processable organolead chloride-bromide mixed halide perovskites for optoelectronic applications. Nano Lett. 2015; 15: 6095.

[75] Zhang, T., et al. A facile solvothermal growth of single crystal mixed halide perovskite $CH_3NH_3Pb(Br_{1-x}Clx)_3$. Chem Commun. 2015; 51: 7820.

[76] Nagane, S., et al. $CH_3NH_3PbI_{(3-x)}(BF4)_x$: molecular ion substituted hybrid perovskite. Chem Commun. 2014; 50: 9741.

[77] Jiang, Q., et al. Pseudohalide-induced moisture tolerance in perovskite CH_3NH_3Pb-$(SCN)_2I$ thin films. Angew Chem. 2015; 54: 11006.

[78] Xue, Q., et al. Metallohalide perovskite—polymer composite film for hybrid planar heterojunction solar cells. RSC Adv. 2015; 5: 775.

[79] Wang, K., et al. p-type Mesoscopic nickel oxide/organometallic perovskite heterojunction solar cells. Sci Rep. 2014; 4: 4756.

[80] Kim, J. H., et al. High-performance and environmentally stable planar heterojunction perovskite solar cells based on a solution-processed copper-doped nickel oxide hole-transporting layer. Adv Mater. 2015; 27: 695.

[81] Jung, J. W., Chueh, C. & Jen, A. K. Y., A low-temperature, solution-processable, Cu-doped nickel oxide hole-transporting layer via the combustion method for high-performance thin-film perovskite solar cells. Adv Mater. 2015; 27: 7874.

[82] Chen, W., et al. Efficient and stable large-area perovskite solar cells with inorganic charge extraction layers. Science. 2015; 350: 944.

[83] Feng, S. L., et al. High-performance perovskite solar cells engineered by an ammonia modified graphene oxide interfacial layer. ACS Appl Mater Inter, in press, DOI: 10.1021/acsami.6b02064.

[84] Wu, Z., et al. Efficient planar heterojunction perovskite solar cells employing graphene oxide as hole conductor. Nanoscale. 2014; 6: 10505.

[85] Christians, J. A., Fung, R. C. M. & Kamat, P. V., An inorganic hole conductor for organo-lead halide perovskite solar cells. Improved hole conductivity with copper iodide. J Am Chem Soc. 2014; 136: 758.

[86] Ganesan, P., et al. A simple spiro-type hole transporting material for efficient perovskite solar cells. Energy Environ Sci. 2015; 8: 1986.

[87] Li, M., et al. Novel spiro-based hole transporting materials for efficient perovskite solar cells. Chem Commun. 2015; 51: 15518.

[88] Burschka, J., et al. Tris(2-(1H-pyrazol-1-yl)pyridine)cobalt(III) as p-type dopant for organic semiconductors and its application in highly efficient solid-state dye-sensitized solar cells. J Am Chem Soc. 2011; 133: 18042.

[89] Burschka, J., et al. Sequential deposition as a route to high-performance perovskite-sensitized solar cells. Nature. 2013; 499: 316.

[90] Jung, M., Raga, S. R., Ono, L. K. & Qi, Y. Substantial improvement of perovskite solar cells stability by pinhole-free hole transport layer with doping engineering. Sci Rep. 2015; 5: 9863.

[91] Chen, W., et al. Simple planar perovskite solar cells with a dopant-free benzodithiophene conjugated polymer as hole transporting material. J Mater Chem C. 2015; 3: 10070.

[92] Franckevičius, M., et al. A dopant-free spirobi [cyclopenta [2,1-b:3,4-b′] dithiophene] based hole-transport material for efficient perovskite solar cells. Mate Horiz. 2015; 2: 613.

[93] Song, Y., et al. Energy level tuning of TPB-based hole-transporting materials for highly efficient perovskite solar cells. Chem Commun. 2014; 50: 15239.

[94] Kazim, S., et al. A dopant free linear acene derivative as a hole transport material for perovskite pigmented solar cells. Energy Environ Sci. 2015; 8: 1816.

[95] Chen, L., Hong, Z., Li, G. & Yang, Y. Recent progress in polymer solar cells: manipulation of polymer: fullerene morphology and the formation of efficient inverted polymer solar cells. Adv Mater. 2009; 21: 1434.

[96] Choi, H., et al. Conjugated polyelectrolyte hole transport layer for inverted-type perovskite solar cells. Nat Commun. 2015; 6: 7348.

[97] Kim, T., Yang, S. & Park, C. Carbon nanomaterials in organic photovoltaic cells. Carbon Lett. 2011; 12: 194.

[98] Heo, J. H., et al. Efficient inorganic–organic hybrid heterojunction solar cells containing perovskite compound and polymeric hole conductors. Nat Photon. 2013; 7: 487.

[99] Bi, D., Yang, L., Boschloo, G., Hagfeldt, A. & Johansson, E. M. J., Effect of different hole transport materials on recombination in $CH_3NH_3PbI_3$ perovskite-sensitized mesoscopic solar cells. J Phys Chem Lett. 2013; 4: 1532.

[100] Kwon, Y. S., Lim, J., Yun, H., Kim, Y. & Park, T. A diketopyrrolopyrrole-containing hole transporting conjugated polymer for use in efficient stable organic–inorganic hybrid solar cells based on a perovskite. Energy Environ Sci. 2014; 7: 1454.

[101] Chiang, T., et al. Functional p-type, polymerized organic electrode interlayer in $CH_3NH_3PbI_3$ perovskite/fullerene planar heterojunction hybrid solar cells. ACS Appl Mater Interface. 2015; 7: 24973.

Chapter 7

First-Principles Study of the Electron Transport Properties of Graphene-Like 2D Materials

Hui Li, Yi Zhou and Jichen Dong

Additional information is available at the end of the chapter

http://dx.doi.org/10.5772/64109

Abstract

First-principles density functional theory and non-equilibrium Green function calculations have been conducted to explore the electronic properties of the graphene-like 2D materials. It is found that zigzag boron phosphide nanoribbons (zBPNRs) exhibit non-magnetic direct bandgap semiconducting property and bandgap is about 1 eV. The heterostructure zSiC-BP-SiC nanoribbons are found to display an obvious negative differential resistance (NDR), which are tunable by changing the length of BPNRs. It is also found that for the armchair MoS_2/WS_2NRs heterostructures, with the number of WS_2NR unit cell increasing, the NDR effect can be modulated. Especially for the case of M(edge) with W atoms doping on the edges, it not only exhibits a significant NDR effect but also owns a fast current transport. Therefore, these graphene-like 2D materials may possess potential for the application in logic transistor.

Keywords: density functional theory, non-equilibrium Green function, BP nanoribbons, MoS_2 nanoribbons, heterostructure, nanoelectronics

1. Introduction

Due to the unique electronic properties and peculiar structure of graphene, such as the zero bandgap and its massless charge carriers, graphene-based nanoscale electronics and photonic devices have been attracting increasing attention [1–7]. Many researchers focused on graphene field effect transistors (FETs) as a result that FETs are the fundamental unit of the integrated circuits. Most studies have presented that graphene shows an exceptionally high carrier mobility, which enables it to easily fabricate into fast-speed FETs [8–12]. However, pristine graphene cannot been applied in the logic transistors as it does not possess a bandgap, not enabling it to gain a large on/off ratio [10, 13, 14]. Therefore, for graphene, opening a sizable bandgap becomes an urgent issue, which should to be tackled by both theoretical and experimental studies. Although considerable methods have been used to solve this problem, such as chemical decorations [15–18], applying a high electric field to bilayer graphene [19–22], and cutting graphene into graphene nanoribbons (GNRs) [23–25], these methods add complexity and depress mobility. To tackle this issue, seeking other graphene-like semiconducting materials would be a better choice.

Nowadays, graphene-like materials have gained considerable research interest because they own significant advantages in low-dimensional scientific research and in nanodevice applications. In 2012, in Alejandro work, they predicted the existence of boron phosphide (BP) nanoflakes, which exhibit extraordinary electronic properties [26]. Interestingly, zigzag SiC nanoribbons (zSiCNRs) share a very similar lattice constant with zigzag BP nanoribbons (zBPNRs), but their electronic properties are different from each other. More recently, 2D transition metal dichalcogenides (TMDs), which are composed of atomic layers coupled by van der Waals forces, have gained considerable interest because of the various electronic properties that they can exhibit, for example, metallic, semiconductor, superconductor, and charge density wave [27–34]. Single-layer MoS_2, which is a semiconductor in a member of TMD family, is not like graphene but have a large bandgap. Hence, in the application of field effect transistor (FET), the MoS_2 nanoflake is a most promising candidate [35, 36]. MoS_2 monolayer consists of a Mo atom layer, which is enclosed within two S layers with hexagonal structure and these layers are held together by weakly coupled van der Waals interactions [37]. Due to such sandwich layer Mo-S-Mo structure, MoS_2 single layer could be probably fabricated repeatedly one by one from bulk materials by micromechanical cleavage [38] or exfoliation [39]. Recently, MoS_2 nanoribbons and other TMD materials with a width down to several nanometers have been synthesized [40–43]. Importantly, among these TMDs, WS_2 is expected to have similar properties with MoS_2 [44–47], and the first experimental demonstrations of monolayer WS_2 have just been reported [48]. Armchair MoS_2 and WS_2 nanoribbons have the same crystal structure and share the very similar lattice constants. Therefore, these nanoribbons might be utilized to construct heterostructures with stable coherent interfaces and multifunctionality.

2. Nanoelectronics constructed using boron phosphide and silicon carbide nanoribbons

2.1. Simulation method

All calculations were performed by the first-principles density functional theory (DFT) and the non-equilibrium Green function method with the Atomistix ToolKit software package [49, 50]. Norm-conserving pseudopotentials were used together with the double-zeta single polarized basis sets. Perdew-Burke-Ernzerhof generalized gradient approximations were employed to describe the exchange-correlation energy. The mesh cutoff was 150 Re. Monkhorst-Pack k-point was set to $1 \times 1 \times 50$ to calculate the Brillouin zone integration. The electron temperature of 300 K was used. The supercells of all structures were built large that vacuum layers between neighboring cells were set to at least 15 Å in order to avoid interactions between periodic images. The geometries of the devices and of the zBPNR crystal structures were optimized until the forces of the atoms were less than 0.05 and 0.01 eV Å^{-1}, respectively. The tolerance for energy convergence was 10^{-} eV. The device current I_d was calculated by the Landauer-Büttiker equation [49]:

$$I = \frac{2e}{h} \int_{\infty}^{-\infty} dE(T(E,V)(f_1(E) - f_2(E))) \tag{1}$$

where $T(E,V)$ represents the quantum mechanical transmission probability for electrons, $f_{1,2}(E)$ denote the Fermi functions of the source and drain electrodes, and e and h are the electron charge and the Planck constant, respectively.

2.2. Results and discussion

2.2.1. The electronics properties of zBPNRs

Figure 1a presents the structure of the hexagonal BP sheet. After optimization, B-P bond length we calculated was 1.858 Å, which is very close to the previously calculated Si-C bond length (1.787 Å) of the hexagonal SiC sheet [51]. The difference between them is less than 5%, so constructing hybrid BP-SiC structures with coherent interfaces was possible. As shown in **Figure 1c**, hexagonal BP sheet is a direct bandgap semiconductor with its highest occupied valence band maximum (HOVBM) and lowest unoccupied conductance band minimum (LUCBM) both at the k-point in the Brillouin zone. P and B atoms π states were observed both in the Bloch states of the HOVBM and the LUCBM, respectively (insets of **Figure 1a**). Moreover, it is also found that the bandgap is 0.87 eV, which is very suitable for electronic applications in FETs.

Figure 1b presents the structure of zigzag BP nanoribbons with the edges passivated by hydrogen. These nanoribbons are denoted as NzBPNR, in which N is the number of zigzag BP chains along the y-axis. Not like zigzag BN nanoribbons with indirect bandgaps [52], zBPNRs still maintain a direct bandgaps after cutting into nanoribbons, where the HOVBM and LUCBM are both located at the K_M point in the Brillouin zone (**Figure 1d**). **Figure 1d** also shows that both the HOVB and the LUCB have a noticeable flat component between the K_M point and the K_Z point, different from others. We can see from **Figure 2a** that the HOVBM of the 6zBPNR locates only at the P atoms, the eigenstates of which diminishes from the P to B edge gradually(left panel), while the HOVB at the K_Z point only locates at the P atoms of P edge. Likewise, as shown in **Figure 2b**, the LUCBM and the LUCB at the K_Z point shares the same trend but the eigenstates locates at the B atoms. Therefore, their edge states should have a great effect on the bandgaps of zBPNRs calculated from HOVBM and LUCBM.

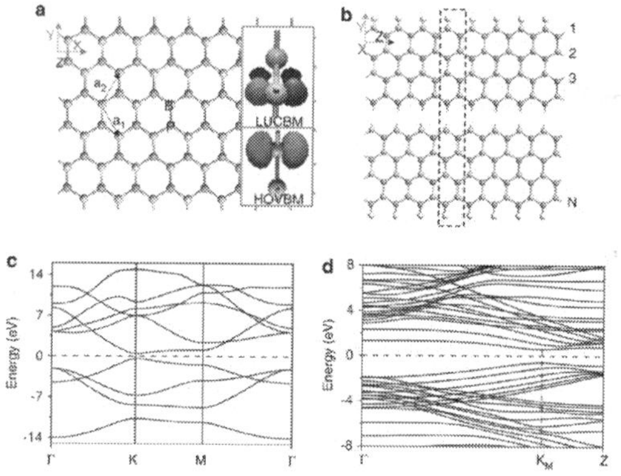

Figure 1. Structures of (a) the hexagonal boron phosphide (BP) sheet and (b) the zigzag boron phosphide nanoribbons (zBPNRs). The arrows and the dashed box denote the unit cell. Band structures of (c) the BP sheet and (d) the 6zBPNR. The light color represents B atom and another one is P atom.

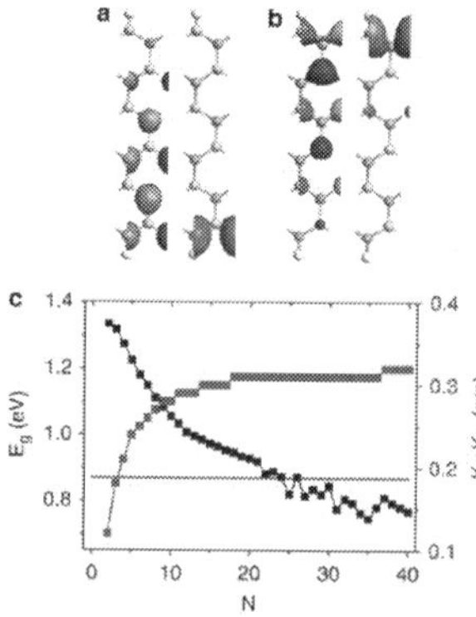

Figure 2. The Bloch states (a) at the highest occupied valence band maximum (HOVBM; left panel) and the HOVBK_Z (right panel), and (b) at the lowest unoccupied conductance band minimum (LUCBM; left panel) and the LUCBK_Z (right panel). The isovalue is 0.1. (c) The variation in the bandgap and the K_z - K^M of zigzag boron phosphide nanoribbons (zBPNRs) with the increase of the ribbon width.

To understand whether the width of the ribbon has effect on the bandgaps (E_g) of zBPNRs, we study the relationship between E_g and N, as shown in **Figure 2c**. Evidently, the E_g dropped monotonously as the ribbon width N increased from 2 to 21. Further increasing N to 40, the E_g

http://dx.doi.org/10.5772/64109

decreases slowly and displays the trend of fluctuation. Moreover, compared to the bandgaps of the BP sheet, when N exceeded 24, E_g of zBPNRs became lower than that which cannot be explained by the quantum confinement effect. As a matter of fact, the above-mentioned phenomena can be explained by two factors, one of which is the electrons transfer between B and P atoms, and the other is the screening effect from the inner part of the ribbon. Because of the different electronegativities of B and P, an electric field and the screening effect can be formed spontaneously when the electron transfers between the two edges of the zBPNRs. As shown in **Figure 2c**, when N increased, the distance between the K_M point and the K_Z point increases and fewer electrons were accumulated at the edges. Therefore, the energy at the B edge decreased and the energy at the P edge increased simultaneously, resulting in the decrease of bandgaps. As the width of the ribbon was sufficiently large and N exceeds 24, we can see from **Figure 2c** that the bandgap fluctuates because of the screening effect emerging to compete with the electron transfer effect. From the above discussion, it can be concluded that zBPNRs may possess potential for applications in optotronics and logic electronics.

2.2.2. *Hybrid zSiC-BP-SiC nanoribbon devices*

Previous studies have shown that the ferrimagnetic state of the 6zSiCNR is a semiconductor with a very small bandgap, while the non-magnetic state of the 6zSiCNR is also a semiconductor with a negligible bandgap and the energy difference between the states of them is very small [53]. Meanwhile, the spin-polarized state would become unstable when external fields are applied [54]. Therefore, we employ a spin-unpolarized calculation to investigate the electron transport properties of the in-plane zigzag SiC and BP(zSiC-BP-SiC) nanoribbon heterostructures. **Figure 3a** shows a structure of such heterostructures, which consists of the scattering region, a hybrid zSiC-BP-SiC structure, and source and drain electrodes, denoted

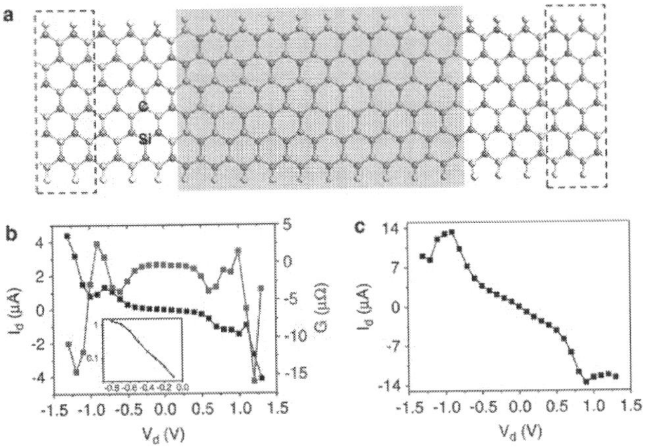

Figure 3. (a) The structure of a zSiC-BP-SiC nanoribbon heterostructure. (b) I_d-V_d and conductance curves of this heterostructure. The inset shows the lgI_d-V_d curve. (c) I_d-V_d curve of a zSiC-BP-SiC nanoribbon heterostructure with a shorter zigzag BP nanoribbon (zBPNR).

by dashed boxes. We study the current-voltage characteristics of this heterostructure, as shown in **Figure 3b**. An obvious non-linear feature was found, due to the discrete energy levels of the nanomaterial. Furthermore, from the I_e-V_d curve, we can see that this heterostructure exhibited a significant conductance gap, resulting from the semiconducting property of the zigzag BP nanoribbons. More interestingly, an obvious NDR effect appeared asymmetrically in the I_d-V_d curve, being at the bias of −0.8 and 1 V, respectively. The asymmetry of the current-voltage characteristics was obviously caused by the asymmetric structure of the scattering region. Further analysis showed that NDR effect could be affected by the width of the BP nanoribbons.

To understand the physical mechanism appearing at the electron transport properties of this heterostructure, transmission spectrum and local density of states (LDOS) [55] at different voltage should be calculated, as shown in **Figure 4**. The LDOS was averaged in the *x*-*y* plane and colored with red and blue, indicating large and small density areas, respectively. In **Figure 4a**, we can see that when there was no bias, the corresponding transmission spectrum exhibited an obvious transmission energy gap, about 0.7 eV around the Fermi level, which was consistent with the current-voltage curves and came from the semiconducting property of the zBPNR, suggesting that such heterostructures have potential applications in FETs [56]. Moreover, a very small transmission peak appeared at the E_f, as shown in the inset of **Figure 4a**, which played an important role in the NDR effect as mentioned above. This small transmission peak was generated by the quantum tunneling effect of electrons between the two zSiCNRs which were located at the two sides of the zBPNR. As shown in **Figure 4d**, the local DOS at the E_f mainly originated from the region of zSiCNR and the SiC-BP interfaces, while there was no electron state at the zBPNR region. As a result, electrons at the E_f were only

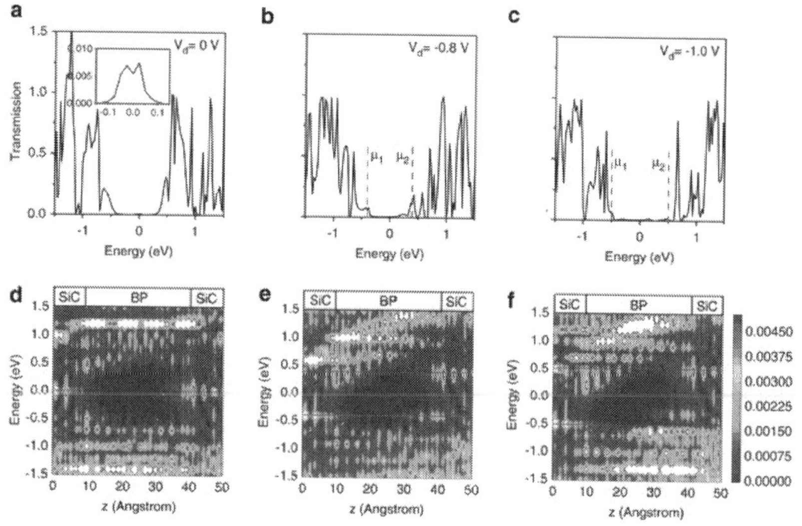

Figure 4. (a-c) Transmission spectrum and (d-f) the local density of states of the heterostructure as shown in **Figure 3(a)** at the bias of 0, −0.8, and −1.0 V, respectively.

able to penetrate the scattering region via the tunneling effect. When the bias decreased from 0 to −0.8 V, as shown in **Figure 4b**,the transmission spectrum went into the bias window, and in the mean time, the transmission of electrons by the tunneling effect at −0.4 eV was increased, which was demonstrated by the LDOS in **Figure 4e** that the states at that energy was enhanced. The transmission spectrum changes in the energy window resulted in a dramatic increase of the current, as shown in the inset of **Figure 3b**. Further decreasing the voltage to −1.0 V, the electron transmission spectrum in the energy window inclined (**Figure 4c**) and the NDR effect appeared, which could be explained that the electron transmission mechanism began to transform from pure tunneling effect to resonant tunneling effect. We can deduce from the above analysis that it is possible to control the tunneling effect. And the NDR effect could also be changed by changing the width of zBPNR. As indicated in **Figure 3c**, the current-voltage curve of a zSiC-BP-SiC nanoribbon heterostructure with a shorter zBPNR consisting of six periodic units clearly showed that the tunneling current was considerably larger than the one discussed above with a longer zBPNR. More importantly, an obvious NDR effect could be also observed in this heterostructure with a shorter zBPNR, and the bias of the current peak asymmetrically located at ±0.9 V, which indicated that the NDR effect of such zSiC-BP-SiC heterostructure could be controlled by the width of zBPNR. As a matter of fact, C_{60}-based structures also exhibited a similar NDR effect, but these structures are much more difficult to be fabricated and may become unstable owing to complicated interface coupling [57].

2.2.3. *Diode-like structure constructed using zBP-SiC nanoribbons*

We also studied the electron transport properties of a diode-like structure consisting of the zigzag BP and SiC nanoribbons, denoted as zBP-SiC. The current-voltage curve of a 5zBP-SiCdiode-like device was shown in **Figure 5**. Due to the semiconducting property of the zigzag BP nanoribbons, we can see there was no current appearing until the bias of ±0.5 V in the current-voltage curve. In addition, when the bias of ±1 V was applied, 5zBP-SiC diode-like device exhibited a slight NDR effect. We can see from the transmission spectrum and the LDOS under different bias that such NDR effect also originated from the change in coupling between the zigzag BP and SiC nanoribbons, as shown in **Figure 6**. As a result, in **Figure 6e**, as the

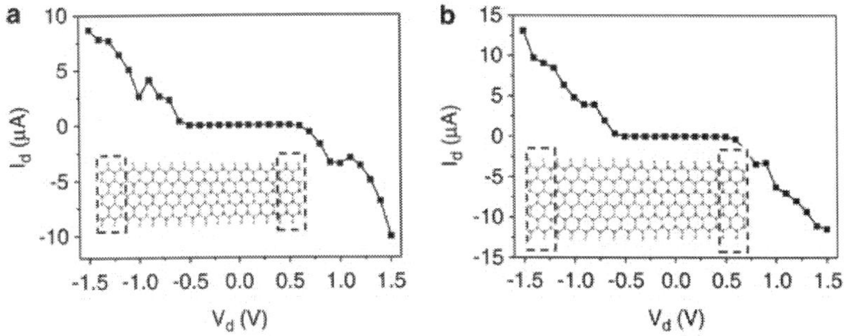

Figure 5. (a) I_d-V_d curve of the 5zBP-SiC two-probe device, and (b) I_d-V_d curve of the 6zBP-SiC two-probe device. The insets are the structures of these devices.

applied bias is −0.9 V, electrons of the lowest unoccupied molecular orbital of the zBPNR contributed to the increase of the current and the electrons of SiC flowed through both the highest occupied molecular orbital and the lowest unoccupied molecular orbital. However, when further decreasing the bias to −1.0 V, we can see from **Figure 6c** and e that the contribution of the highest occupied molecular orbital of the zSiCNR moving away from the bias window and the current was significantly depressed, leading to the decrease of the current in **Figure 5a**. In **Figure 5b**, we also study the current-voltage curve of this diode-like device with a wider width, which showed that NDR effect weakened and disappeared as the width of the ribbon increased.

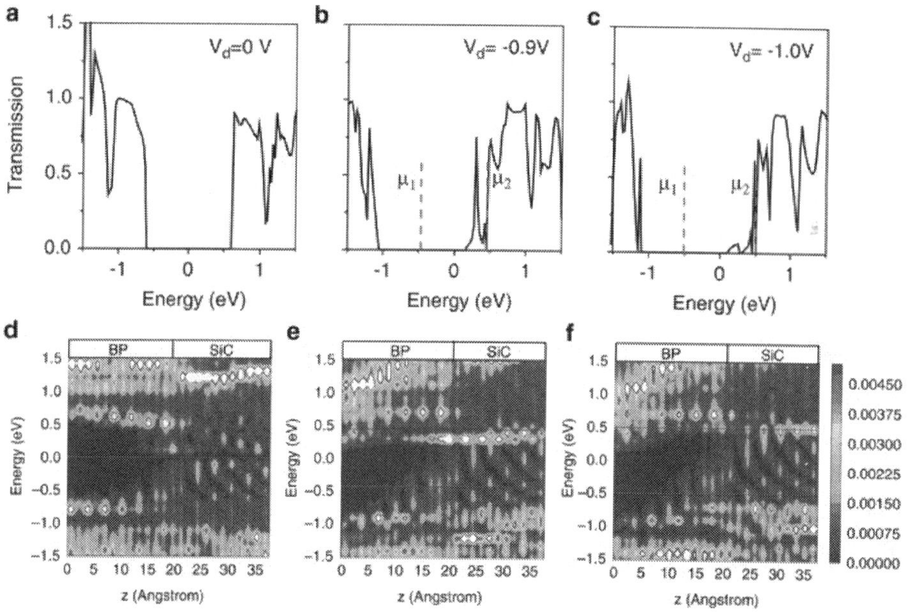

Figure 6. (a-c) Transmission spectra, and (d-f) local density of states of the two-probe device shown in **Figure 5(a)** at V_d of 0, −0.9, and −1.0 V, respectively.

To investigate whether the zBPNRs played an important role in the heterostructures mentioned above, we investigate the electron transport properties of folded 4zSiCNRs, as shown in **Figure 7**, with the overlapping area decreasing from device a to d. The transmission spectrum of the four cases indicates that there was no transmission energy gaps near the Fermi energy, because the non-magnetic 4zSiCNR is a narrow-gap semiconductor. As mentioned above, it is concluded that only the zSiCNR structure is difficult to construct into nanoelectronics.

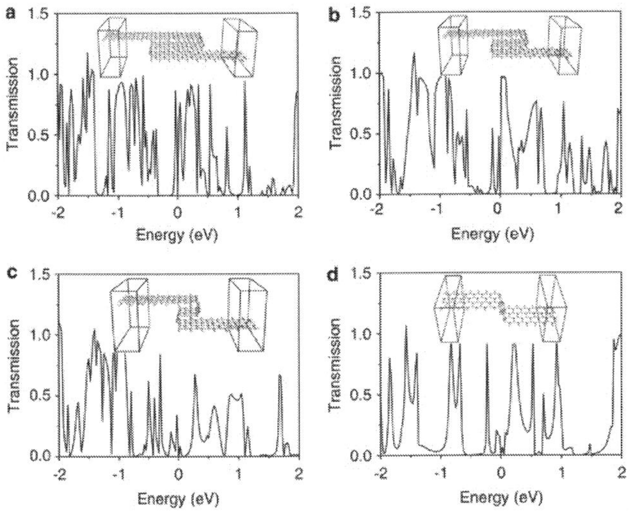

Figure 7. Transmission spectra of fourfold 4zSiCNR (silicon carbide nanoribbon) two-probe devices. The insets illustrate the corresponding structures of these devices.

3. In-plane heterostructures constructed using MoS_2 and WS_2 nanoribbons

3.1. Simulation method

All calculations were performed by the first-principles density functional theory (DFT) and the non-equilibrium Green function method with the Atomistix ToolKit software package [49, 50]. Norm-conserving pseudopotentials were used together with the double-zeta single polarized basis sets. Perdew-Burke-Ernzerhof generalized gradient approximations were employed to describe the exchange-correlation energy. The mesh cutoff was 150 Re. Monkhorst-Pack k-point was set to 1 × 1 × 100 to calculate the Brillouin zone integration. The electron temperature was employed by 300 K. The supercells of all structures were built large that vacuum layers between neighboring cells were set to at least 15 Å in order to avoid interactions between periodic images. The geometries of the devices and of the crystal structures were optimized until the forces of the atoms were less than 0.05 and 0.01 eV $Å^{-1}$, respectively. The tolerance for energy convergence was 10^{-5} eV. The device current d_d was calculated by the Landauer-Büttiker equation [49]:

$$I = \frac{2e}{h} \int_{\infty}^{-\infty} dE(T(E,V)(f_1(E) - f_2(E))) \tag{2}$$

where $T(E,V)$ represents the quantum mechanical transmission probability for electrons, $f_{1,2}(E)$ denote the Fermi functions of the source and drain electrodes, and E and h are the electron charge and the Planck constant, respectively.

3.2. Results and discussion

Figure 8 shows the devices consisting of an armchair MoS_2NRs/WS_2NRs heterostructure. The devices are placed along the z direction, and the x-axis is perpendicular to its surface. The two-probe device consists of two parts, which are the scattering region, making of MoS_2NRs/WS_2NRs heterostructure, and source and drain electrodes, the supercells of which are denoted in yellow box. And these heterostructures are described by M(na) and M(nz), where n indicates the width of the WS_2NR in the scattering region, and a(z) denotes the armchair or zigzag-shaped ribbon.

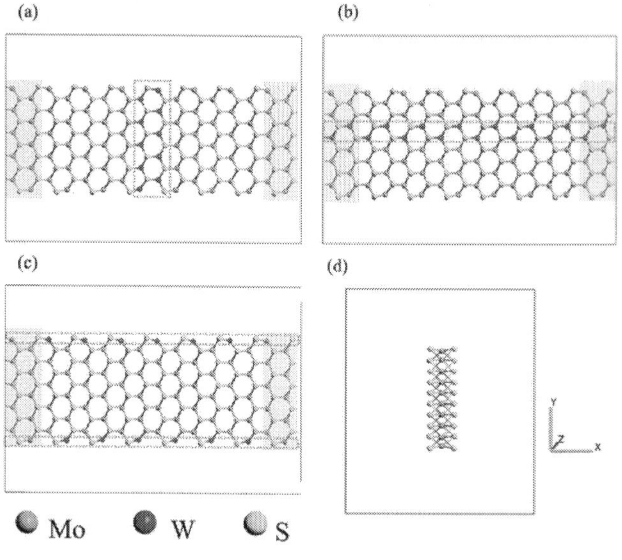

Figure 8. Structure of MoS_2NR/WS_2NR heterostructure two-probe devices. Dashed box denotes the contacting WS_2NR. The supercells of the devices are marked by solid rectangle. (a) Structure of M(1a) two-probe device. (b) Structure of M(1z) two-probe device. (c) Structure of M(edge) two-probe device. (d) Side view of these heterostructures.

3.2.1. Heterostructure of WS_2/MoS_2 with zigzag direction interface

In Gong et al. [58] work, they found that in-plane heterostructure of zigzag direction interface between WS_2 and MoS_2could be mostly appeared in the MoS_2/WS_2 hybrid structures. Therefore, we first construct M(na) with the interfaces along the zigzag directions, in other words, MoS_2NRs and WS_2NRs are arranged in perpendicular. And the electron transport properties of these heterostructures are investigated. The current-voltage curves of M(na) are presented in **Figure 9** with WS_2NRs lengths *n* ranging from 1 to 3. It can be clearly seen that no matter the length of the WS_2NRs, all the three cases exhibit an obvious conductance gap in the current-voltage curves before the bias of 0.6V, which results from the semiconducting property of the armchair MoS_2 nanoribbon. Moreover, on further increasing the bias, these heterostructures all exhibit a significant NDR effect, which locate at 0.9 V bias. As indicated from **Figure 9**, the WS_2NR length has impact on the NDR effect, with the length of WS_2NR increasing, the current is enhanced more quickly and M(3a) owns the most current peak among these two-probe devices. In addition, peak-to-valley ratio (PVR) that represents the extent of the NDR effect also increases, namely 3.01, 4.931, and 5.441, respectively, and the NDR window are the same, that is, 0.1 V. It is concluded that it exhibits a relatively good NDR performance when contacting more unit cells of WS_2NRs. From the above discussion, it indicates that these heterostructures may possess the potential for application of logic transistor.

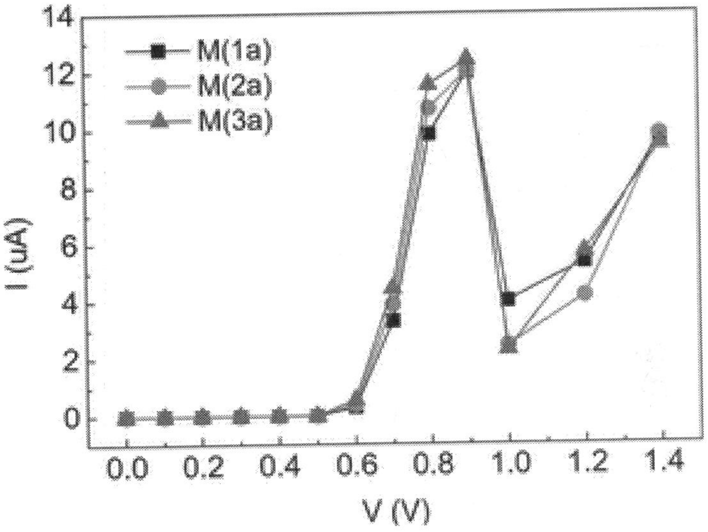

Figure 9. I_d-V_d curves of M(*n*a) with WS_2NR length *n* ranging from 1 to 3.

We now investigate the physical origin of the NDR; the transmission spectrum is shown in **Figure 10(a)** for M(3a) at several typical bias voltages. When a small bias of 0.6 V is applied, a little transmission peak occurs in the transmission energy window and spontaneously the current begins to appear initially. Further increasing the bias to 0.9 V, the transmission peak under the bias window broadens, resulting in the increase of current and up to the maximum current. Interestingly, as applied the bias to 1.0 V, the height of transmission peak drops largely, which play the most dominant contribution to the current. Although there are additional two small peaks appearing in the bias window, the negative contribution to the current can be ignored. Therefore, the NDR effect can be found in the **Figure 8**. And when the bias continuously increases to 1.4 V, the transmission peaks under the expanding bias window increase and broaden, which results in a steady increase in current. In **Figure 10(b)** and **(c)**, to explain whether the NDR effect of these three heterostructures is different from each other, transmission spectrum at bias of 0.9 and 1.0 V is calculated. In the case of the applied bias of 0.9 V, all the cases have the single peak under the bias window. However, when the length of the WS_2NR enlarges, the height of transmission peak also increases but slowly, such that the current peak of these cases has a small increase, while for the cases of 1.0 V bias, with n increasing from 1 to 3, the majority of peaks under the bias window decrease. However, there is another small transmission peak exhibiting an opposite trend, but the contribution to the current is ignored. Therefore, when the length of the WS_2NR decreases, the current valley decreases simultaneously in **Figure 9**. And from **Figure 10(b)**, we can also investigate the reduction extent which is consistent with variation of the current-voltage curves that as the length of the WS_2NR n increases from 1 to 2, the degree of reduction of transmission peaks denoted by b and c performs more significantly than when it increases from 2 to 3.

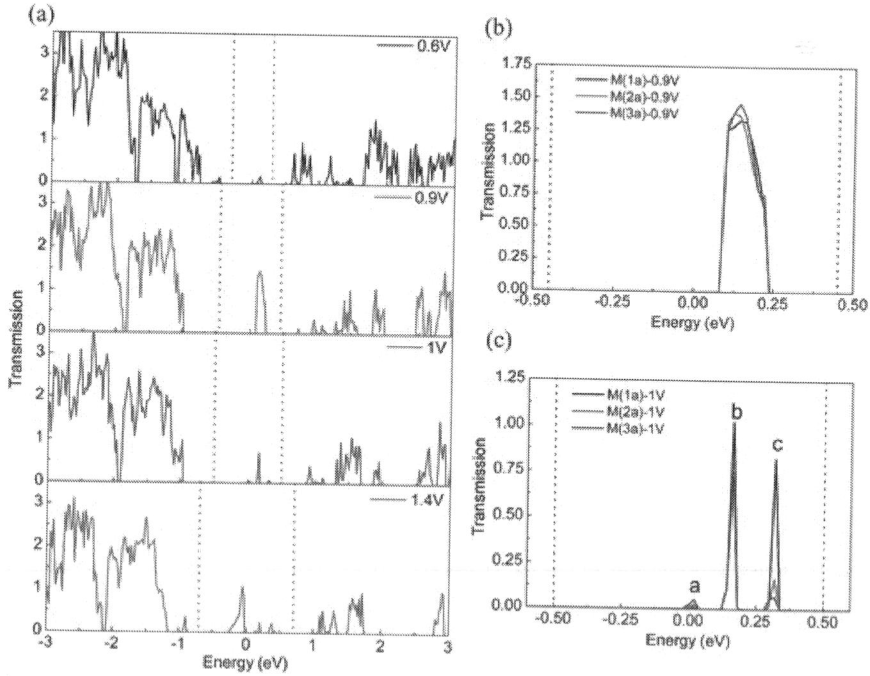

Figure 10. Transmission spectrum (a) for M(3a) at four typical biases, and for M(na) (b) at the bias of 0.9 V and (c) at the bias of 1.0 V as the WS_2NR length n ranges from 1 to 3. Dotted line represents the bias window.

http://dx.doi.org/10.5772/64109

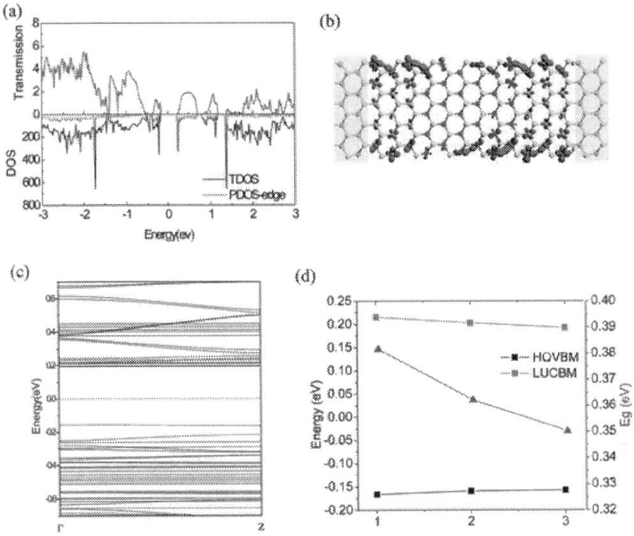

Figure 11. (a) Equilibrium transmission spectrum, and total and projected density of states for M(3a), (b) LDOS at E_f −0.18 eV with an isovalue of 0.1, (c) band structures for M(3a), and (d) HOVBM, LUCBM, and bandgap for M(*n*a) as the WS_2NR length *n* ranges from 1 to 3.

To identify the nature states of the electron transport properties, equilibrium density of states (DOS) of these devices and their corresponding band structure of crystal structures are calculated and shown in **Figure 11**. In **Figure 11(a)**, the curve of DOS indicates that zero electron states occur at E_f because of the semiconductor characteristics of armchair MoS_2 and WS_2 nanoribbons, not resulting in electron transmission at E_f, which correspond to current-voltage curves (**Figure 9**) that there is a conductance gap before the bias of 0.6 V. In addition, the changes of transmission peaks around E_f can be explained directly by the DOS. Interestingly, the projected density of states on the edge (PDOS edge) is mainly consistent with the total density of states (TDOS) in the energy range from −0.4 to 1.2 eV, indicating that the states from the edges contribute to the electrons through the scattering region near E_f. Besides, to describe this phenomenon clearly, we also calculate the local density states (LDOS) at −0.18 eV, as shown in **Figure 11(b)**. It can be clearly seen that at that energy, the electron states originate from the W and Mo atoms on the two edges. Moreover, as the WS_2NR length increases, the electron states are enhanced and becomes bigger and bigger. From the above discussion, it can be concluded that electrons are prone to transport the scattering region from the edges of transition metal atoms with a higher energy, which can be explained by the effect of dangling bonds. From the analysis of the band structure, we can see that armchair MoS_2/WS_2NRs heterostructures remain as a direct bandgap semiconductor with its highest occupied valence band maximum (HOVBM) and lowest unoccupied conductance band minimum (LUCBM) both at Γ point in the Brillouin zone, as shown in **Figure 11(c)**, which indicates that the good optical characteristics of MoS_2 retain in the armchair MoS_2/WS_2NRs heterostructures. The above discussion means that MoS_2/WS_2NRs heterostructures may possess potential for their application in optotronic devices. The information regarding the changes of LUCBM and HOVBM among these heterostructures is shown in **Figure 11(d)**. With the length of WS_2NRs increasing, the bandgaps decline, which results from the decrease of LUCBM and the increase of HOVBM. Although there are changes in the bandgaps of these heterostructures, when the bias is applied, the current increases more quickly with a higher current peak in **Figure 9**. The variation of the bandgap we observed is consistent with Gong et al. [58] work that the PL peak position shifted continuously across the interface.

3.2.2. *Heterostructure of WS_2/MoS_2 with armchair direction interface*

In addition, as Gong et al. [58] work mentioned, besides the preferred zigzag interface, in-plane heterostructures of $WS_2/MoS_2$with the armchair interface were also occasionally observed. Therefore, we also study the electron transport properties of such armchair MoS_2/WS_2NR heterostructures with the interface along the armchair direction with the increasing length of WS_2NR. In **Figure 12**, the current-voltage curves of M(*n*z) with *n* ranging from 1 to 4 are presented. Apparently, these heterostructures all exhibit significant NDR effect, located at the applied bias from 0.9 to 1.0 V. Moreover, in the case of M(4z), another NDR effect is observed at the bias of 1.2 V with an ignored PVR. All these heterostructures own the same NDR window, that is, 0.1 V. When the width of WS_2NR increasing from 1 to 4, the NDR effect is depressed and PVR is 3.3037, 2.7585, 1.6063, and 1.003, respectively. Especially for M(4z), the NDR effect is the most inferior, but the current increases more significantly than the other cases. From the above discussion, it is concluded that these hybrid two-probe devices may not be a good candidate for the application in logic transistor.

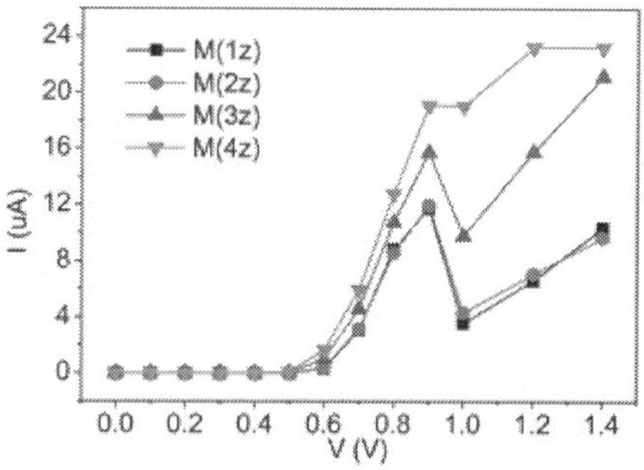

Figure 12. I_d-V_d curves of M(*n*z) with WS_2NR length *n* ranging from 1 to 4.

To understand the observed NDR effect, it is useful to analyze the evolution of the transmission spectrum as the bias potential is ramped up for M(1z) and some typical biases are shown in **Figure 13(a)**. When a small bias of 0.6 V is applied, transmission peak begins to appear in the bias window, which results in an increase in the current. Under the bias of 0.9 V, only one transmission peak appeared in the bias window and peak height gradually enhances as the bias is applied. Simultaneously, the weight of the transmission spectrum in the bias window is enlarged, leading to the increase of current. When the bias is 1.0 V, an additional transmission peak appears in the expanding bias window, but the only peak in the bias window of 0.9 eV drops significantly. Compared to each other, the height of peak reduction outweighs another small peak appearing in the bias window, resulting in the decrease of current. As a result, current-voltage curves exhibit the NDR effect (**Figure 12**). When further increasing the bias to 1.4 V, the transmission peaks in the bias window broaden and there are some small peaks moving into the extending bias window, leading to a steady increase in the current. For the case of M(4z), with a wider WS_2NR, transmission spectrum at some several typical biases is presented in **Figure 13(b)**. It is more obvious that the initial transmission peak appears in the bias window and with the bias increasing, the height of the peak enhances. When the bias exceeds 0.9 V, the contribution of transmission spectrum both at the bias of 1.0 and 1.4 V under the bias window all decrease slightly, leading to twice NDR effect with the ignored PVR. In **Figure 13(c)** and **(d)**, we study the distinction of these four heterostructures, and transmission spectra at bias of 0.9 and 1.0 V are calculated. When the width of WS_2NR increases from 1 to 4, the transmission peaks both at two biases broaden and therefore both the current peak and valley increase. After investigating the variation of PVR, we find that as the width of WS_2NR broadens, the increasing extent for bias of 1.0 V is more obvious than the bias of 0.9 V. From the above discussion, it is explained the reason why the NDR effect is depressed.

To investigate the mechanism of the electron transport properties, we calculate the equilibrium TDOS and PDOS on the edges of M(1z), as shown in **Figure 14(a)**. It can be seen that a significant electron state gap appear around E_f in the DOS, which results from the semiconducting feature of armchair MoS_2 nanoribbons. Moreover, it directly corresponds to the transmission spectrum that there is also an obvious transmission gap occurring. Interestingly, the projected density of states on the edge (PDOS-edge) is mainly consistent with the TDOS around Fermi level, indicating that the states from the edges contribute to the electrons through the scattering region. Besides, to describe this phenomenon clearly, we also calculate the local density states (LDOS) at −0.18 eV, as given in **Figure 14(b)**; it can be clearly seen that at the energy of −0.18 eV, the electron states originate from the Mo atoms on the two edges. In other words, electrons transport through the scattering region mainly from the transition metal atoms of the edge. As discussed earlier, because the Mo atoms on the edges have higher energy with unfilled d orbit, the electron can easily transport from them. Moreover, band structure of the crystal is analyzed in **Figure 14(c)**. From the analysis of the band structure, we can see that these four heterostructures exhibit a direct bandgap semiconducting property with HOVBM and LUCBM both at Γ point in the Brillouin zone, as shown in **Figure 14(c)**. The optical characteristics of armchair MoS_2 nanoribbons do not change after being in contact with WS_2, and these hybrid two-probe devices may also possess the potential application in optotronics. To clearly investigate the variation of the bandgaps among these heterostructures, we extract the information in **Figure 14(d)**. When the width of WS_2NR increases, there is a slight decrease in bandgap, resulting from the decrease of LUCBM and the increase of HOVBM. Therefore, more electrons can jump from HOVBM to LUCBM easily and as the bias is applied, the current increases more quickly with a higher current peak. Moreover, Gong et al. [58] work confirms our finding that the PL peak position shifted continuously across the interface.

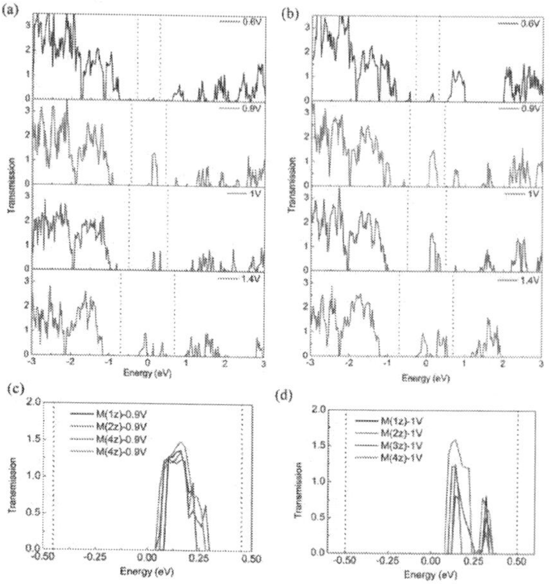

Figure 13. Transmission spectra for (a) M(1z) and (b) M(4z) at four typical biases. Dotted line represents the bias window. Transmission spectra for M(nz) (c) at 0.9 V bias and (d) at 1.0 V bias as width of WS_2NR n ranges from 1 to 4.

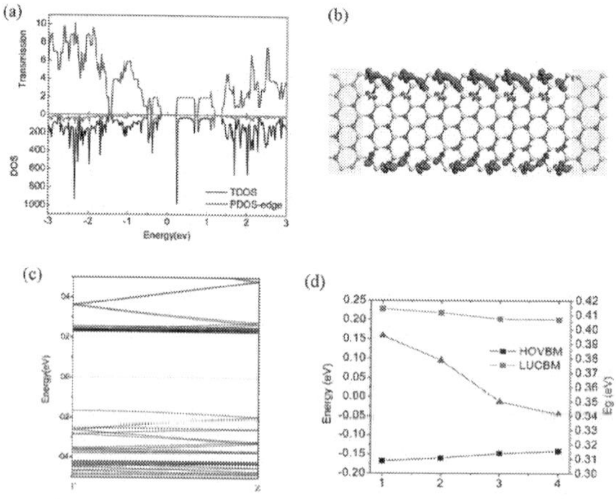

Figure 14. (a) Equilibrium transmission spectrum, total and projected DOS, (b) LDOS at E_f −0.18 eV with an isovalue of 0.1, (c) band structures of M(1z), and (d) HOVBM, LUCBM, and bandgap for M(nz) as the length of WS_2NR n ranges from 1 to 4.

3.2.3. MoS_2/WS_2 heterostructure with W atoms doping on the edges

Lastly, we study the another heterostructure denoted by M(edge) that Gong et al. [58] did not discuss, in which the Mo atoms on the edges are replaced by W atoms. Due to the same amount of W doping, M(1z) makes a comparison with this heterostructure to investigate whether edge states influence electronic transport properties. In **Figure 15**, we calculate current-voltage curves for M(edge). Compared to M(1z), a significant NDR effect also appears in M(edge), but the NDR window enlarges to 0.3 V. When the bias exceeds 0.6 V, an initial current appears and increases more rapidly than the case of M(1z), and then the current continues to enhance up to its maximum at 0.9 V. Although the current peak is almost similar to the case of M(4z), there is a lowest current valley among these heterostructures, which leads to the best NDR effect and largest PVR is 18.4462. Interestingly, M(edge) and M(1z) are doped the same amount of W atoms. However, the M(edge) performs more excellently than M(1z) and exhibits a better NDR effect and the faster current transport. In fact, this phenomenon can be explained that the W atoms on the edge possess the higher energy, resulting in the electrons of W atoms easily transporting through the edges than from the inside. From the previous discussion, M(edge) exhibits not only a significant NDR effect but also a fast current transport, which would make it the best candidate in the application of logic transistor.

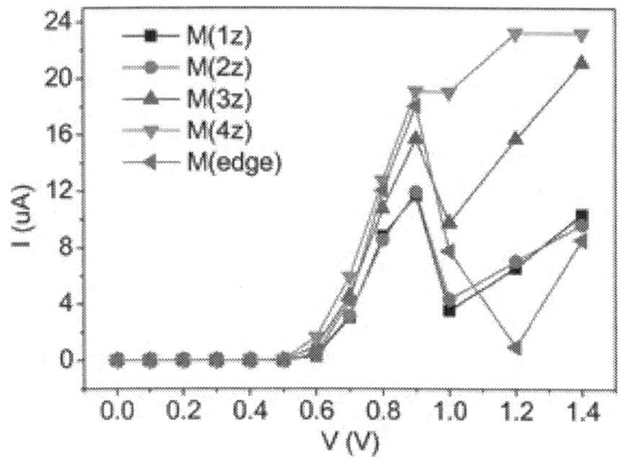

Figure 15. I_d-V_d curves of M(edge) and M(nz) with WS_2NR length *n* ranging from 1 to 4.

To understand the observed NDR effect, transmission spectra at four typical biases are calculated, as shown in **Figure 16(a)**. When applied a small bias of 0.6 V, the transmission energy window occurs a little transmission peak, resulting in an initial current. When the bias is 0.9 V, the single transmission peak in the bias window broadens up to its maximum. Further increasing the bias, this only transmission peak reduces obviously into two small peaks. Therefore, it displays a significant NDR effect. To clearly investigate the distinction between the M(1z) and M(edge), we calculate the transmission spectrum at the bias of 0.9, 1.0, and 1.2 V, respectively, as shown in **Figure 16(b-d)**. For the bias of 0.9 and 1.0 V, the transmission spectrum of M(edge) contributes larger than M(1z) in the bias window and results in a higher current. However, in the case of 1.2 V bias, for M(edge), the height of two peaks under the bias window are both lower, resulting in a small current.

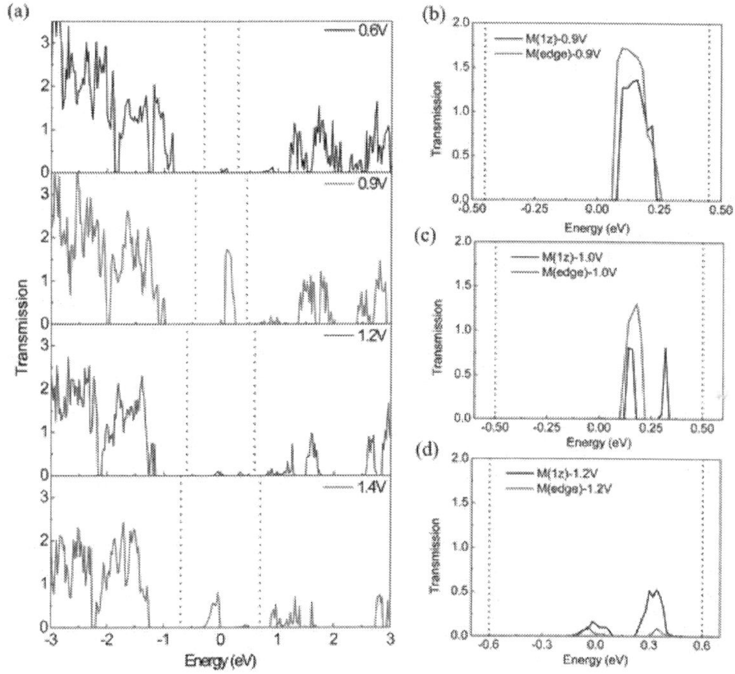

Figure 16. (a) Transmission spectrum for M(edge) at four bias voltages. Dotted line represents the bias window. Transmission spectrum (b) at 0.9 V bias, (c) at 1.0 V bias and (d) at 1.2 V bias for M(1z) and M(edge).

4. Conclusion

First-principles DFT and non-equilibrium Green function calculations have been used to study the electronic properties of the graphene-like 2D materials. It was found that zBPNRs exhibited a non-magnetic direct bandgap semiconducting property and bandgap was about 1 eV. We also found that when the width of zBPNRs increases, the bandgap decreases below the level of BP monolayer, resulting from electrons transport from the P edge to the B edge of the zBPNRs. Moreover, a heterostructure, which consists of a zBPNR and two zSiCNRs, was constructed and the electron transport property was studied. For these zSiC-BP-SiC two-probe devices, a significant NDR effect was observed, arising from the change in the SiC-BP coupling under various biases, and when changing the length of zBPNR, the NDR effect of these heterostructures can be modulated. In addition, we also studied the electron transport properties of a diode-like structure consisting of the zigzag BP and SiC nanoribbon. For these zBP-SiC two-probe devices, the NDR effect weaken with the increasing width of ribbon, and zBPNRs played an important role in these heterostructures.

For all the armchair MoS_2/WS_2NRs heterostructures, they are found to be a direct bandgap semiconductor. When enlarging the width of WS_2NR, bandgap of these heterostructures narrows slightly, which leads to the current increasing fast and owning a higher current peak. Moreover, when the width of the WS_2NR increases, for the case of M(na) with the interfaces along the zigzag directions, the NDR effect becomes a little better, while for M(nz) with the interfaces along the armchair directions, the NDR effect becomes inferior but with a higher current peak. Interestingly, for M(edge) with W atoms doping on the edges, it not only exhibits a significant NDR effect but also a fast current transport. Therefore, M(edge) may possess the great potential for the application in logic transistor.

Author details

Hui Li[1*], Yi Zhou[1] and Jichen Dong[2]

*Address all correspondence to: lihuilmy@hotmail.com

1 Key Laboratory for Liquid-Solid Structural Evolution and Processing of Materials, Ministry of Education, Shandong University, Shandong, China

2 Department of Mechanical and Biomedical Engineering, City University of Hong Kong, Kowloon Tong, Hong Kong

References

[1] Ritter, K. A., Lyding, J. W. The influence of edge structure on the electronic properties of graphene quantum dots and nanoribbons. Nat. Mater. 2009;8:235–242.

[2] Shen, L., Zeng, M. G., Yang, S. W., Zhang, C., Wang, X. F., Feng,Y. P. Electron transport properties of atomic carbon nanowires between graphene electrodes. J. Am. Chem. Soc. 2010;132:11481–11486.

[3] Mativetsky, J. M., Liscio, A., Treossi, E., Orgiu, E., Zanelli, A.,Samorì, P., et al. Graphene transistors via in situ voltage-induced reduction of graphene-oxide under ambient conditions . J. Am. Chem. Soc. 2011;133:14320–14326.

[4] Westervelt, R. M. Graphene nanoelectronics. Science. 2008;320:324–325.

[5] Novoselov, K. S., Geim, A. K., Morozov, S. V., Jiang, D., Zhang, Y., Dubonos, S. V., et al. Electric field effect in atomically thin carbon films. Science. 2004;306:666–669.

[6] Chen, J. -H., Jang, C., Xiao, S., Ishigami, M., Fuhrer, M. S. Intrinsic and extrinsic performance limits of graphene devices on SiO_2. Nat. Nanotech. 2008;3:206-209.

[7] Chen, F., Xia, J., Ferry, D. K., Tao, N. Dielectric screening enhanced performance in graphene FET. Nano Lett. 2009;9:2571–2574.

[8] Lin, Y. M., Dimitrakopoulos, C., Jenkins, K. A., Farmer, D. B., Chiu, H. Y., Grill, A., et al. 100-GHz transistors from wafer-scale epitaxial graphene. Science. 2010;327:662–662.

[9] Hollander, M. J., LaBella, M., Hughes, Z. R., Zhu, M., Trumbull, K. A., Cavalero, R., et al. Enhanced transport and transistor performance with oxide seeded high-κ gate dielectrics on wafer-scale epitaxial graphene. Nano Lett. 2011;11:3601–3607.

[10] Liao, L., Bai, J. W., Qu, Y. Q., Lin, Y. C., Li, Y. J., Huang, Y., et al. High-K oxide nanoribbons as gate dielectrics for high mobility topgated graphene transistors. Proc. Natl. Acad. Sci. U.S.A. 2001;107:6711–6715. [14] 6

[11] Elias, D. C., Gorbachev, R.V., Mayorov, A.S., Zhukov, A.A., Blake, P., Ponomarenko, L.A., et al. Dirac cones reshaped by interaction effects in suspended graphene. Nat. Phys. 2011;7:701–704.

[12] Mayorov, A. S., Gorbachev, R.V., Morozov, S.V., Britnell, L., Jalil, R., Ponomarenko, L.A., et al. Micrometer-scale ballistic transport in encapsulated graphene at room temperature. Nano Lett. 2011;11:2396–2399.

[13] Castro Neto, A. H., Guinea, F., Peres, N. M. R., Novoselov, K. S., Geim, A. K. The electronic properties of graphene. Rev. Mod. Phys. 2009;81:109–162. 2

[14] Li, X., Cai, W. W., An, J. H., Kim, S. Y., Nah, J., Yang, D. X., et al. Large-area synthesis of high-quality and uniform graphene films on copper foils. Science. 2009;324:1312–1314. 7

[15] Balog, R., Jørgensen, B., Nilsson, L., Andersen, M., Rienks, E., Bianchi, M., et al. Bandgap opening in graphene induced by patterned hydrogen adsorption. Nat. Mater. 2010;9:315–319.

[16] Gan, L., Zhou, J., Ke, F., Gu, H., Li, D. N., Hu, Z. H., et al. Tuning the properties of graphene using a reversible gas-phase reaction. NPG Asia Mater. 2012;4(11):e31

[17] Zhang, W. J., Lin, C. T., Liu, K. K., Tite, T., Su, C. Y., Chang, C. H., et al. Opening an electrical band gap of bilayer graphene with molecular doping. ACS Nano. 2011;5:7517–7524.

[18] Dong, J.C., Li, H. Monoatomic layer electronic constructed by granphene and boron nitride nanoribbons. J. Phys. Chem. 2012;116:17259–17267.

[19] Zhang, Y., Tang, T.-T., Girit, C., Hao, Z., Martin, M.C., Zettl, A., et al. Direct observation of a widely tunable bandgap in bilayer graphene. Nature. 2009;459:820–823.

[20] Castro, E. V., Novoselov, K. S., Morozov, S. V., Peres, N. M. R., Lopes dos Santos, J. M. B., Nilsson, J., et al. Biased bilayer graphene: semiconductor with a gap tunable by the electric field effect. Phys. Rev. Lett. 2007;99:216802–216805.

[21] Ohta, T., Bostwick, A., Seyller, T., Horn, K., Rotenberg, E. Controlling the electronic structure of bilayer graphene. Science. 2006;313:951–954.

[22] Zhang, Y. B., Tang, T. T., Girit, C., Hao, Z., Martin, M. C., Zettl, A., et al. Direct observation of a widely tunable bandgap in bilayer graphene. Nature. 2009;459:820–823.

[23] Li, X., Wang, X., Zhang, L., Lee, S., Dai, H. Chemically derived, ultrasmooth graphene nanoribbon semiconductors. Science. 2008;319:1229–1232.

[24] Son, Y. W., Cohen, M. L., Louie, S. G. Energy gaps in graphene nanoribbons. Phys. Rev. Lett. 2006;97:216803–216806.

[25] Hicks, J., Tejeda, A., Taleb-Ibrahimi, A., Nevius, M. S., Wang, F., Shepperd, K., et al. A widebandgap metal-semiconductor-metal nanostructure made entirely from graphene. Nat. Phys. 2013;9:49–54.

[26] Alejandro, L. C. Prediction of boron-phosphorous nanographene-like material. Int. J. Quantum Chem. 2012;112:3152–3157.

[27] Zhang, H., Li, X.-B., Liu, L.-M. Tunable electronic and magnetic properties of WS2 nanoribbons. J. Appl. Phys. 2013;114:093710.

[28] Chhowalla, M., Shin, H. S., Eda, G., Li, L.-J., Loh, K. P., Zhang, H. The chemistry of two-dimensional layered transition metal dichalcogenide nanosheets. Nat. Chem. 2013;5:263.

[29] Zhang, H., Liu, L.-M., Lau, W.-M. Dimension-dependent phase transition and magnetic properties of VS_2. J. Mater. Chem. A. 2013;1:10821.

[30] Wilson, J. A., Yoffe, A. D. The transition metal dichalcogenides discussion and interpretation of the observed optical, electrical and structural properties. Adv. Phys. 1969;18:193–335.

[31] Fivaz, R., Mooser, E. Mobility of charge carriers in semiconducting layer structures. Phys. Rev. 1967;163:743–755.

[32] Radisavljevic, B., Radenovic, A., Brivio, J., Giacometti, V., Kis, A. A. Single-layer MoS_2. Nat. Nanotechnol. 2011;6:147–150.

[33] Ye, J. T., Zhang, Y. J., Akashi, R., Bahramy, M. S., Arita, R., Iwasa, Y. Superconducting dome in a gate-tuned band insulator. Science. 2012;338:1193–1196.

[34] Withers, R. L., Wilson, J. A. An examination of the formation and characteristics of charge-density waves in inorganic materials with special reference to the two and one-dimensional transition-metal chalcogenides. J. Phys. C. 1986;19:4809–4845.

[35] Liu, H., Neal, A. T., Ye, P. D. Channel length scaling of MoS_2 MOSFETs. ACS Nano. 2012;6:8563–8569.

[36] Radisavljevic, B., Kis, A. A. Mobility engineering and a metal–insulator transition in monolayer MoS_2. Nat. Mater. 2013;12:815–820.

[37] Verble, J. L., Wietling, T. J., Reed, P. R. Rigid-layer lattice vibrations and Van der Waals bonding in hexagonal MoS2. Solid State Commun. 1972;11:941–944.

[38] Novoselov, K. S., Jiang, D., Schedin, F., Booth, T. J., Khotkevich, V. V., Morozov, S. V., et al. Two-dimensional atomic crystals. Proc. Natl. Acad. Sci. U.S.A. 2005;102:10451–10453.

[39] Coleman, J. N., Lotya, M., O'Neill, A., Bergin, S. D., King, P. J., Khan, U., et al. Two-dimensional nanosheets produced by liquid exfoliation of layered materials. Science. 2011;331:568–571.

[40] Wang, Z., Li, H., Liu, Z., Shi, Z., Lu, J., Suenaga, K., et al. Mixed low-dimensional nanomaterial: 2D ultranarrow MoS_2 inorganic nanoribbons encapsulated in quasi-1D carbon nanotubes. J. Am. Chem. Soc. 2010;132:13840–13847.

[41] Liu, X., Xu, T., Wu, X., Zhang, Z., Yu, J., Qiu, H., et al. Top–down fabrication of sub-nanometre semiconducting nanoribbons derived from molybdenum disulfide sheets. Nat. Commun. 2013;4:1776.

[42] Liu, Z., Suenaga, K., Wang, Z., Shi, Z., Okunishi, E., Iijima, S. Identification of active atomic defects in a monolayered tungsten disulphide nanoribbon. Nat. Commun. 2013;2:213.

[43] Nethravathi, C., Anto Jeffery, A., Rajamathi, M., Kawamoto, N., Tenne, R., Golberg, D., et al. Chemical unzipping of WS2 nanotubes. ACS Nano. 2013;7:7311–7317.

[44] Zhu, Z. Y., Cheng, Y. C., Schwingenschlögl, U. Giant spin-orbit-induced spin splitting in two-dimensional transition-metal dichalcogenide semiconductors. Phys. Rev. B. 2011;84:153402.

[45] Jiang, H. Electronic band structures of molybdenum and tungsten dichalcogenides by the GW approach. J. Phys. Chem. C. 2012;116:7664.

[46] Ramasubramaniam, A. Large excitonic effects in monolayers of molybdenum and tungsten dichalcogenides. Phys. Rev. B. 2012;86:115409.

[47] Feng, W., Yao, Y., Zhu, W., Zhou, J., Yao, W., Xiao, D. Intrinsic spin Hall effect in monolayers of group-VI dichalcogenides: a first-principles study. Phys. Rev. B. 2012;86:165108.

[48] Gutierrez, H.R., Perea-López, N., Elías, A.L., Berkdemir, A., Wang, B., Lv, R., et al. Extraordinary room-temperature photoluminescence in triangular WS2 monolayers. Nano Lett. 2013;13:3447–3454.

[49] Brandbyge, M., Mozos, J. L., Ordejon, P., Taylor, J., Stokbro, K. Density-functional method for nonequilibrium electron transport. Phys. Rev. B. 2002;65:165401–165417.

[50] ATK. QuantumWise. Copenhagen, Denmark. www.quantumwise.com [Internet]. 2008.

[51] Sun, L., Li, Y. F., Li, Z. Y., Li, Q. X., Zhou, Z., Chen, Z. F., et al. Electronic structures of SiC nanoribbons. J. Chem. Phys. 2008;129:174114–174117.

[52] Park, C. H., Louie, S. G. Energy gaps and stark effect in boron nitride nanoribbons. Nano Lett. 2008;8:2200–2203.

[53] Lou, P., Lee, J. Y. Band structures of narrow zigzag silicon carbon nanoribbons. J. Phys. Chem. C. 2009;113:12637–12640.

[54] Li, Z. Y., Qian, H. Y., Wu, J., Gu, B. L., Duan, W. H. Role of symmetry in the transport properties of graphene nanoribbons under bias. Phys. Rev. Lett. 2008;100:206802.

[55] Lu, W. C., Meunier, V., Bernholc, J. Nonequilibrium quantum transport properties of organic molecules on silicon. Phys. Rev. Lett. 2005;95:206805.

[56] Qi, J. S., Huang, J. Y., Feng, J., Shi, D. N., Li, J. The possibility of chemically inert, graphene-based all-carbon electronic devices with 0.8 eV gap. ACS Nano. 2011;5:3475–3482.

[57] Zheng, X. H., Lu, W. C., Abtew, T. A., Meunier, V., Bernholc, J. Negative differential resistance in C60-based electronic devices. ACS Nano. 2010;4:7205–7210.

[58] Gong, Y., Lin, J., Wang, X., Shi, G., Lei, S., Lin, Z., et al. Vertical and in-plane heterostructures from WS2/MoS2 monolayers. Nat. Mater. 2014;13:1135–1142.